THE INFRASTRUCTURE BOOK

How Cities Work and Power Our Lives

SYBIL DERRIBLE

Prometheus Books

Essex, Connecticut

Prometheus Books

An imprint of The Globe Pequot Publishing Group, Inc.
64 South Main Street
Essex, CT 06426
www.globepequot.com

Distributed by NATIONAL BOOK NETWORK

British Library Cataloguing in Publication Information available

Library of Congress Cataloging-in-Publication Data available
ISBN 9781493086641 (paperback)
ISBN 9781493086658 (electronic)

♾️™ The paper used in this publication meets the minimum requirements of American National Standard for Information Sciences—Permanence of Paper for Printed Library Materials, ANSI/NISO Z39.48-1992

To Marie-Agathe,
my partner in travel,
my partner in life.

CONTENTS

What Is Infrastructure and How to Read This Book

What you are doing right now, reading these lines, whether you are holding a physical book or using an electronic device, would not be possible without the essential infrastructure that surrounds us and enables our lives.

Our society has become so reliant on the seamless and silent operation of infrastructure that it is hard to imagine a time when that infrastructure did not exist (and it still does not exist in some parts of the world). In fact, modern infrastructure has such a ubiquitous presence in our society that we easily forget it is even there. This is the infrastructure that is at the center of *The Infrastructure Book*.

I often say that infrastructure is the greatest physical manifestation of civilization. All over the world, and since we started to settle together in cities, humans have built infrastructure to engineer the environment they live in and make it more favorable to them. Already about ten thousand years before our common era (BCE), during the Neolithic period, humans built roads, canals, and buildings to make life easier for them. In *Politics*, Aristotle, who lived in the fourth century BCE, provides this insight: "The city-state comes into being for the sake of living, but it exists for the sake of living well." This aspect of "living well" is fundamental, and it explains why cities form and grow in the first place. It also explains why so much effort and energy is put into building and operating gigantic infrastructure systems that have become so big and complex that it is impossible to understand fully how they work as a single entity.[1]

---

1. In software engineering, fixing a bug in a code can easily create more bugs because of the interconnections between all pieces of the software. It is the same thing with infrastructure. The networks of infrastructure systems have become so complex that it has become impossible to forecast fully the

What is mind-boggling is that even though our lives depend on infrastructure—and when it breaks down we are sent back to the dark ages—most of us have little to no idea how it works. We expect light to come on when we flip a switch, and we expect water to flow when we open a tap. But do you know where your electricity comes from and how it is generated? Do you know where your water comes from and how it is treated and distributed? We are very good at cursing and complaining when we are stuck in traffic, but do you really know what traffic congestion is and how to ease it? We are amazing at consuming stuff and throwing waste in the dumpster, but do you know how it is managed and where it ends up after the garbage truck picks it up? And when you make a call, watch television, or browse the Web on your computer, phone, or tablet, do you know how the information is transmitted, seamlessly, instantaneously?

This last point is important. After tremendous technological progress in the nineteenth and twentieth centuries, the twenty-first century seems to be the century of the *instant*. Everything is instantaneous. In many countries, people can wash instantly by turning a knob instead of having to fill up buckets of water at a well. Water can be drunk readily from the tap; it does not need to be boiled first. Light comes on at the flip of a switch, and the use of candles has become reserved for romantic gestures. Most often, we do not even need to think about space heating and cooling; it is all taken care of automatically for our comfort. Even in terms of knowledge, thanks to the internet, most people have instantaneous access to the universal body of knowledge that is stored on servers around the world and reaches users at light speed. Everything has become so instantaneous that many people now feel the need to travel to the countryside, to disconnect, but to disconnect from what? How does the infrastructure that provides this instantaneous life work in the first place?

Having instantaneous access to these services also comes at a great cost. Planning, designing, building, and operating all this infrastructure

---

ripple effects that a change in one system has on itself and on other systems. We have massive models that can come up with reasonable estimates, but not with a complete and accurate forecast.

requires a tremendous amount of energy and resources, often more than what the Earth can provide sustainably, at least as of this writing. Moreover, the sheer complexity of infrastructure has also become so great, and our lives depend so much on it functioning well, that this comfort we have gained has come at the expense of resilience. In the news it seems that there is always a region of the world that is experiencing an extreme weather event, destroying cities and infrastructure and catastrophically impacting people's lives. Sustainability and resilience are two important concepts that we will discuss throughout the book, as they require us to rethink completely what infrastructure is and how it works.

Funnily, despite being a societal necessity, the term *infrastructure* itself is relatively new—it only first appeared in France in the 1870s. As the combination of *infra* (meaning "below") and *structure*, *infrastructure* was originally defined as "an underlying base or foundation especially for an organization or system." By the twenty-first century, the use of the term *infrastructure* has become so broad that it is impossible to come up with a proper and complete definition. Infrastructure is most often associated with *public utilities* (whether publicly or privately owned), and that is the view I will take in this book. Specifically, we will discover seven types of infrastructure: water, wastewater, transport, electricity, gas, solid waste, and telecommunication. Our focus will be on the planning and operation of specific infrastructure, whether physical (e.g., water conduits) or an offered service (e.g., public transport). Along the way, we will also identify their biggest flaws and discuss how they are or may evolve to become more sustainable and resilient.

Broadly, my goal is to provide a general understanding of the *engineering principles* that govern the seven infrastructure systems listed above. These engineering principles often impose constraints on what can and cannot be done in our communities. Understanding them can also help us figure out how we can become more sustainable and resilient. We will cover many engineering principles but will not use any equations (although I list some in the footnotes for those interested).

You do not require any prior engineering knowledge to read this book. In fact, I use this book to teach first-year university students, many of whom have no engineering knowledge and despise math. As I tell my students, after reading this book, you should be able to walk in on any technical conversation about infrastructure and get the gist of it. You should be able to correct that cranky person who complains that there is too much traffic and claims they have the solution for it (most often they don't). Most important, you should be able to stand on any street in any city in the world and have some idea of how the infrastructure that surrounds you works and what its merits and flaws are. I promise you that by the time you are finished with this book, you will not see the world around you the same way ever again.

Before we start, we must have a discussion that some readers will find thorny. There is no engineering without numbers, and there are no numbers without units. I use only the metric unit system since this is the system used internationally and often in the industry (even in the United States). For fans of the imperial unit system, I have added its equivalent in parentheses . . . but you know it is only a matter of time before the United States fully converts to the metric unit system. The metric system is simply better and much easier to use.[2]

The seven infrastructure systems listed above are covered over sixteen chapters. For each infrastructure system, we will travel to a different city in the world to illustrate how the engineering principles work in practice. For example, we will travel to Hong Kong to learn about water distribution, Copenhagen to learn about stormwater management, Los Angeles to learn about traffic and roads, Cape Town to learn about electricity generation, and San Francisco to learn about digital telecommunication. Finally, we will travel to Singapore to summarize everything we have learned. Except for this chapter on Singapore, each chapter is independent enough that you should not have to read them sequentially (although I recommend you do).

---

2. In my classes, I use only the metric unit system. When I started teaching in the United States, I told students I would include imperial units too, but they asked me not to. They did not like the imperial system. Conversions between units like feet to miles are not obvious. The metric system is much easier and more natural to use.

In addition to the sixteen cities that we will travel to, a second infrastructure is featured at the end of each chapter to see how the engineering principles that we will learn about work in different contexts. For example, we will also learn about Bogotá's TransMilenio at the end of the public transport chapter in London and about Oaxaca's solution to mobile telecommunication at the end of the analog telecommunication chapter in New York City.

Are you ready? We are about to embark on an amazing journey together to explore seven essential infrastructure systems. A journey around the world. A journey that will probably transform your understanding of infrastructure and give you a new appreciation for it.

Our first stop is Rome. That's where we will learn about water collection. Let's go.

Part One

WATER

CHAPTER ONE

Water Collection

All Aqueducts Lead to Rome

REGARDLESS OF NATIONALITY, GENDER, RACE, OR WEALTH, EVERYONE needs water to live. Water is as universal a need for human beings as it gets. It is not a surprise that most cities in the world are built next to sizable bodies of water, whether a river or a lake. Think of Chicago located next to Lake Michigan, Baghdad by the Tigris River, Vienna by the Danube River, Kolkata by the Ganges River, and Ho Chi Minh City by the Saigon River; the list goes on and on. In fact, most large cities that are not built close to sizable bodies of water eventually tend to have severe water shortage problems, as Rome experienced once it outgrew what the Tiber (Rome's main river) could supply.

I remember my first trip to Rome in 2011. My girlfriend at the time (who is now my wife) and I had first visited Venice, which left quite an impression on me. We then rented a car in Florence and drove to Rome. Walking around Rome is surprisingly easy. We were amazed by how small the city center was compared to other large European capitals like Paris and London. At every turn, a new monument awaited. It is quite an experience. The density of monuments per square kilometer in Rome must be one of the highest in the world. I remember walking around the Piazza Navona and the Pantheon one evening, in search of the Piè Di Marmo, which a travel guide had recommended as a quirky little piece of history. Piè Di Marmo is a remnant of a statue built in the first century, and the only thing that remains is a left foot. I had no idea what to expect. There it was, at a street corner, a relatively large left foot tucked next to a building, surrounded by a small metal fence. It was more quirky than impressive, but I remember thinking that this single small piece of a statue must have seen so much history, from its early Roman days through the Italian Renaissance, witnessing the rise of the automobile, all the way to the present day. It is amusing to try to imagine what that unassuming left foot must have gone through over the course of its existence.[1]

In the Western world, Rome is often credited as being the first city to have had a population of one million people. What is astounding is that this threshold was achieved around the year 1 CE (common era), when there was no access to electricity, motorized vehicles, or any other infrastructure that our modern society thrives on and that we will learn about in this book. Personally, no matter how much I try, I simply cannot grasp how life must have been under first-century conditions. What the Romans missed in infrastructure, however, they made up for with great engineering skills.

When a city is relatively small, getting water directly from a nearby water source (like a river) or digging a few wells into the ground is enough. But as the urban population increases, so does the demand for water. When that happens, there are essentially two solutions: reduce the demand or increase the supply. In the first instance, for example, we can reduce or ration the amount of water used per person so that the total amount of water available remains sufficient for everyone. In the second instance, we increase the supply by engineering our way out of the problem thanks to new infrastructure. We will see this type of problem time and again throughout this book, as it is directly linked to sustainability, and we will often find that the right solution to tackle our modern problems is not to engineer our way out of them but to reduce the demand.[2]

---

1. Naturally, I had to see it again the next time I visited Rome.
2. You can read more about it in chapter 2 of my book *Urban Engineering for Sustainability* (Cambridge, MA: MIT Press, 2019) and in the article S. Derrible, "An Approach to Designing Sustainable Urban Infrastructure," *MRS Energy and Sustainability* 5, no. 1 (2019): 13.

However, sustainability was not a global issue at the time of the Romans, and they excelled at engineering their way out of their problems.

We must remember that the Tiber is a relatively small river. It cannot provide enough water for a population of one million. Plus, when urban population increases, local sources of water are bound to get polluted. In Rome, initially, the Tiber, springs, and wells were used as freshwater sources throughout the city,[3] but after some time, the amount of water supplied could not match the amount needed for the growing population. What is the solution? Either you limit the growth of the city and tell people to go live somewhere else or you bring water from somewhere else into the city. The Romans took the second option.

In this chapter, we will learn about water collection as the process of transporting water over long distances before it is to be processed for distribution—that is, before it is to be sent to a pipe network and to individual buildings. In the twenty-first century, water collection systems are mostly used to bring water to water treatment plants. The Romans did not have any significant treatment process, but they brought water from the countryside to *distribution basins*, located on the edges of towns. To bring all this water, they had to master water collection.

To be able to bring water into the city, we must first understand how water flows. It turns out that there is one single governing force when it comes to water flow, a force that can be either our greatest ally or our greatest enemy. Manipulating and mastering this force can change everything and solve many problems. This force is none other than—gravity. Simply put, when it comes to water, gravity will systematically make water flow from a higher elevation to a lower elevation. In other words, if no other forces are exerted, water will always flow downward, never upward. We are taught this phenomenon when we are kids in school, when we learn about the water cycle. Do you remember that illustration? The one with the ocean on one side, the sun above it with the arrows showing

---

3. As Frontinus wrote in his book *The Aqueducts of Rome*, "For four hundred and forty-one years from the foundation of the City, the Romans were satisfied with the use of such waters as they drew from the Tiber, from wells, from springs." Frontinus was a Roman engineer, author, and politician.

evaporation processes, and on the other side water flowing down mountains and rivers, all the way until it reaches the ocean again. This is what gravity does. In fact, regardless of how flat a river might appear when you walk around a city, there is always going to be a slope, if only a tiny tilt, so that one side is lower than the other, and the water will flow toward the lower side. The Romans did not have sophisticated pumps. They had to understand and master gravity.

Conceptually, what the Romans did is simple: they created artificial rivers, commonly called *aqueducts*. Because the Tiber River could not provide enough water, they engineered new rivers to bring more water to the city. They went to the countryside and found freshwater sources that were at higher elevations than Rome. Then they built channels with a downward slope to allow water to flow down to Rome. While the general concept is simple, mastering the engineering skills to build a channel and control its slope is not. They had to dig through mountains and build monumental bridges solely to transport water. These projects took years to complete and were expensive. They include both the well-known elevated structures (such as the famous arcades of the Pont du Gard—look it up if you do not know it) and covered trenches, tunnels, canals, pipes, and other pieces of infrastructure that allow water to flow from its source all the way to its destination.

The first aqueduct built in Rome was Aqua Appia. It was built by Appius Claudius Caecus[4] and completed in 312 BCE. Aqua Appia spanned roughly sixteen kilometers (ten miles), starting east of Rome. The elevation of Aqua Appia at its source was around twenty meters above sea level, and the elevation once in Rome is around 8.5 meters, a difference of 11.5 meters (38 feet). That means that over sixteen kilometers, the Romans managed to engineer an aqueduct by strictly controlling for elevation to maintain an average slope of less than a tenth of a percent.[5] This achievement is incredible for the time. Remember that

---

4. Appius Claudius Caecus was a *censor*, not an emperor; Rome was not an empire at the time.
5. The slope can be approximated as $11.5 / 16{,}000 = 0.0007 = 0.07\%$.

there was no electricity, no trucks, and no laser (used today in geomatics to precisely measure elevation).

Aqua Appia was the first of many aqueducts. At the end of its glory days, by the third century CE, Rome was served by eleven aqueducts, not including all the aqueducts built throughout the Roman Empire, providing enormous volumes of water to its various cities. Most aqueducts were built about one meter underground in covered trenches, following the geography of the land—that is, simply following the natural slopes of the land. Covered trenches were easier to build, and they are less affected by natural elements like the sun and the wind. Moreover, they can seamlessly pass through a city, unlike walls and bridges. Perhaps needless to mention, the initial surveying to identify the best route over which to build the aqueduct was crucial. The preference was to avoid complicated routes that required the construction of tunnels through mountains and bridges over rivers since they are difficult and costly to build.[6]

Of the eleven aqueducts that were built, only one is still working today. Originally named Aqua Virgo, it was renamed Acqua Vergine after it was restored during the Italian Renaissance; Acqua Vergine provides water to the Trevi Fountain and the Piazza del Popolo. Aqua Virgo was the sixth aqueduct built. It was completed in 19 BCE. Aqua Virgo spanned roughly twenty-one kilometers (thirteen miles), and the difference in elevation from the source to the destination was around four meters (thirteen feet), thus representing a slope of two hundredths of a percent.[7]

Small slopes may be more difficult to deal with, but they are preferred. Remember that gravity is the main force acting on the flow—it governs both direction and velocity. Steeper slopes mean that gravity applies a larger force, resulting in a higher flow velocity.[8] A faster flow is

---

6. As a general indication, tunnels and bridges represent about 20 percent of the length of an aqueduct.

7. Here, the slope is 4 / 21,000 = 0.0002 = 0.02%. Such a small slope leaves no room for error.

8. The typical equation used nowadays to estimate the flow Q in an open channel is the Manning equation: $Q = \dfrac{C_m}{n} \cdot A \cdot R^{2/3} \cdot S_f^{1/2}$, where C_m is a conversion coefficient if the metric system is not

more difficult to handle, and the force exerted by the flow[9] on the channel can wear excessively its bottom and sides, eventually leading to greater maintenance needs.

There is one case in which steeper slopes are preferred. At times, water had to cross a valley over which a bridge could not be built, and the mid-point of the valley was lower than the elevation of the point across the valley. In this case, Romans built pipes. Unlike a river that has a surface exposed to the air (the technical term is *open channel*), in a fully enclosed pipe, pressure can build up.[10] If the pressure is high enough, it can create a force greater than gravity so that water can flow upward in the pipe. Imagine a bowl with one side higher than the other. To have water flow down to the bottom of the bowl and then up to the other side, you can use a small tube (or loose straw) and pour water into the tube on the higher side, and the pressure will build up at the bottom and push the water to the other side. You do not need a pump at the bottom; gravity provides the force that increases the pressure in the tube. Technically, this is called an *inverted siphon*; figure 1.1 shows how it works. For valleys, you can do the same thing with a properly sized pipe. But controlling forces in a pressurized pipe is not easy, as we will see when we travel to Hong Kong.

Going back to water collection: in addition to channels and pipes, aqueducts were composed of a multitude of basins serving different functions and connecting the channels and the pipes. *Collection basins* accumulated water at the source before it was sent off to the aqueduct. Throughout the length of the aqueduct, *settling basins* (or *settling tanks*) were installed—these basins were larger and deeper to allow the flow velocity to decrease so that heavier objects like debris could settle at the

---

used, n is the Manning roughness, A is the area of the cross-section of the flow, R is the hydraulic radius (ratio of the area to the wetted perimeter), and S_f is the slope in the water surface (that is where gravity appears). We will not get into the details of the equation, but it is less scary than it looks. What we can see is that a rougher surface (i.e., higher n) means a smaller flow, that a larger area means a higher flow, and that a larger slope also means a higher flow (i.e., a higher velocity thanks to gravity). In the end, controlling the slope of the channel largely governs the flow since this is where gravity expresses itself.

9. Called the shear forces.

10. A partially full pipe is an open channel, like a river.

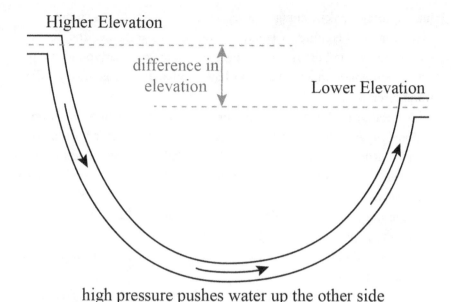

Higher Elevation

difference in
elevation

Lower Elevation

high pressure pushes water up the other side

Figure 1.1. *Inverted siphon.*

bottom while the clean water at the surface could continue its journey down the aqueduct. *Regulating basins* were placed in strategic areas when the amount of water entering the system could be too large, and this way water could be diverted from the main channel; similar pieces of infrastructure were built in Rome to divert excess water to natural water bodies, including the Tiber. Junction and splitting basins were also built to merge incoming channels into one or to split a single channel into multiple channels (e.g., to serve two areas in the city).

If you have visited Roman waterworks, you may be more familiar with *distribution basins.* Distribution basins were large tanks located at the end of aqueducts, whether at the entrance of a town or by a villa. This point is where water supply to the city could be controlled. Through a system of pipes, water from distribution basins flew down to *storage basins,* fountains, bathhouses, and private houses (the latter two for a fee paid to the state). In the form of either reservoirs (open to the air) or cisterns (underground tanks), storage basins acted as a buffer so that more water than what was coming in could be consumed over time. The Basilica Cistern in Istanbul offers a stunning example of an underground storage basin. Nowadays, many cities also have similar basins, generally called water reservoirs.

Overall, the Romans were resourceful. They managed to engineer a regional network of infrastructure to provide enough water for everyone living in Rome. Again, Rome got to a population of one million people. Being able to provide water for that many people is a feat. In the end, so that everyone had access to water, the Romans built and operated about thirteen hundred fountains located throughout the city that were free to use by the population.[11] To this day, Rome is full of *nasoni*—small water fountains—where tourists and locals can fill up their water bottles with tasty tap water.[12]

The Romans were not the first to build water infrastructure. At least in the Western world, several civilizations, including the Greeks, had built significant pieces of water infrastructure such as tunnels and cisterns. Because every human being requires water to survive, it is not surprising to find water infrastructure in every civilization. The need for water and the power of water are so great that water is also often used as a metaphor in local cultures to explain complex ideas. My favorite is the aphorism *panta rhei* by the ancient Greek philosopher Heraclitus, who lived in the fifth century BCE. *Pantha rhei* means "everything flows," expressing that nothing is ever constant in life,[13] a theme that is central to Eastern philosophies as well.

---

11. To learn more about Roman aqueducts, many sources can be found on the Web and in other books; I recommend the website http://www.romanaqueducts.info/ (accessed May 29, 2024), which possesses a wealth of information on the topic.

12. The water fountains are grayish and are the size of fire hydrants. *Nasoni* means "big nose" because the faucet of the water fountains looks like a big nose. A website mentions that Rome has over twenty-five hundred nasoni that were built mostly in the nineteenth century. The water is smooth and crisp and incredibly refreshing. Rome has some of the best tap water I have ever had.

13. The same idea is conveyed by the saying that no one ever steps twice in the same river. The molecules of water that brush your feet when you step in water are never twice the same.

Back to infrastructure: despite not being the first civilization to build water infrastructure, the Romans left an important legacy that has inspired many cities to follow suit, even in modern society. The Catskill Aqueduct has been providing water to New York City since 1924. It is around 150 kilometers long (92 miles), and it uses the same principles as the Roman aqueducts, with 60 percent of the aqueduct being made of covered trenches. In California, an intricate water collection system was built to provide water to Los Angeles from the Owens Valley; dubbed the "California water wars," the fight over the system was fueled by political drama that inspired the movie *Chinatown*. In China, a more-than-2,400-kilometer aqueduct was built in the early 2000s to bring water from the water-rich south to the water-poor north—the project is called the South-North Water Transfer Project. Rome itself follows its own history. Built in the twentieth century, most of the water distributed in Rome today travels through the 130-kilometer-long Peschiera-Capore aqueduct—one of the longest in the world—from the Peschiera and Capore springs.

Another form of water collection is pumping the water that is in the ground (called *groundwater*) from what are called *aquifers* (i.e., sources of groundwater). Of all fresh water on Earth, surface water bodies such as lakes and rivers only contain about half a percent. Icecaps and glaciers contain nearly 70 percent of all fresh water. The remaining 30 percent is held in underground aquifers. Groundwater therefore represents a large source of fresh water for many cities. We will not get too far into the details of groundwater collection, but we should mention that there are two kinds of groundwater aquifers: *unconfined* and *confined*.

Unconfined aquifers are sources of water that are below your feet. Soils have pores into which water trickles, again because of gravity. Moreover, because water is heavier than air, water trickles down until all pores are saturated. The top surface where the soil is saturated with water is called the *water table*. Pumps are installed below the level of the water table—often at the bottom of a well—and water can be pumped and sent to water treatment plants. The bottom of the well consists of solid rock with no or very small pores, making it impractical for water to travel farther down. Beneath the solid rock, however, there can also be soil with pores saturated with water. These are the confined aquifers. Water in confined aquifers is trapped between two layers of solid rock that act like a pipe. As a result, pressure can build up. This is why, at times, when drilling down through solid rock, water shoots up by itself.[14] These types of wells are also called *artesian* wells, after the former province of Artois in France.

The mechanisms between confined and unconfined wells are a little different because of the pressure buildup in confined aquifers, but where they meet is in their limitation. Namely, if pumping rates are above the replenishment rates, the aquifers will dry out. Knowing the properties of an aquifer is therefore critical.

There is one more thing we need to mention about groundwater. Beyond the fact that it is more present on Earth than surface water, many water utilities prefer to collect groundwater because it tends to be cleaner and easier to treat (we will focus on water treatment in the next chapter). Going back one step, we must realize that riverbeds are seldom made of impermeable rock. They are usually porous. Gravity doing its thing means that the soil beneath rivers is saturated with water. In fact, a little like icebergs whose tip only shows a fraction of it, river water is often only a fraction of all the water that is flowing down a valley. Most of it flows beneath the surface. The surface of a river acts as the water table, except that it is exposed to the atmosphere instead of being underground. Many utilities prefer to collect water that is in the ground beneath rivers[15] because it is usually cleaner and easier to treat. Moreover, if you go deep enough, you usually do not have to worry about droughts that can affect the river.

From ancient Roman aqueducts to modern Chinese water transfer projects, water collection infrastructure systems highlight the importance of water in a city. Yet so far we have discussed only water collection to

---

14. This is similar to how oil shoots up, since it also tends to be under intense pressure underground.

15. Horizontal wells are built—that is, wells are drilled next to rivers and then redirected horizontally so the water beneath the river is pumped.

bring water to a city. That is the first step. The second step is to make the water drinkable. Let's travel to Tel Aviv, where people drink water from the seas, to learn about water treatment.

THE AQUEDUCTS OF TOKYO

Tokyo has an amazing history. By 1721, in over a little more than a century, it transformed from a small fishing village to a city housing more than one million people. Although the boom in Tokyo happened later than it did in Rome, the technology available in the 1600s was closer to what the Romans had than what was available to cities by the second half of the twentieth century. Like Rome, Tokyo was built in low-lying areas, meaning it could also leverage gravity to transport fresh water.

Josui is a Japanese word that translates to "drinking waterway" in English. Essentially, *josuis* are manmade canals. One of the *josuis* the Japanese built is forty-three kilometers (twenty-seven miles) long, taking its source from the Tama River, west of the city, and transporting water all the way to Shibuya in the center of the city. On the way, some of the water is also used for irrigation, turning infertile land into fertile land, and helping provide food for everyone.

So far, the Tokyo story resembles Rome's, but it is not over yet. Unlike Rome, through which the Tiber flows, Tokyo does not have one main river where it can dispose of its sanitary sewers (i.e., wastewater generated by humans) that would flow downstream. Tokyo is located at a bay entrance. Any wastewater disposed of in a river would sit in the bay or even wash ashore. What is the solution? What if we could turn this problem into an opportunity?

To most of us, wastewater is gross and disgusting, but it is also full of nutrients and serves as a great fertilizer. The solution in Tokyo was to collect the sanitary wastewater and to sell it to farmers. They would come into the city bringing their produce, and they would leave with buckets of wastewater to serve as fertilizers, again helping turn infertile land into fertile land.

Beyond providing water infrastructure, the canals and rivers of Tokyo were also used as transport infrastructure. In the 1700s, trucks did not exist. The best way to transport freight was by boat. In fact, Tokyo has more than eight hundred kilometers (five hundred miles) of rivers and canals, whose role in history cannot be overemphasized. If you look at Japanese prints from the seventeenth, eighteenth, or nineteenth century that depict life in Tokyo,[16] chances are that you will see many canals. To all readers who have traveled to Tokyo, you may not remember

---

16. Many prints at the time adopted the Ukiyo-e style.

seeing canals spreading across the city. What has happened? Where did the canals go? Modernization happened. The city continued to grow and became industrialized. Industrialization often meant pollution, and the canals indeed became heavily polluted. The popularization of the car also required a lot of space for roads. This need for space led to many canals being infilled and covered with concrete or being culverted (i.e., contained in a drain and paved over). From what I have read, the 1964 Olympic Games acted as a catalyst to modernize the city and get rid of the canals. At the same time, the different wards of Tokyo entered a "garbage war" that we will discuss when we cover solid waste management later in the book.

In the future, some canals may be restored. At least that's my hope. There is something about the sight of water that is calming and reassuring, especially in busy places like Tokyo. Plus, canals can be strategic assets that serve multiple purposes, as we will discover when we visit Seoul.

CHAPTER TWO

Water Treatment

Drinking from the Seas in Tel Aviv

FOR THE LONGEST TIME, THE QUALITY OF THE WATER THAT WAS naturally available to cities was adequately potable. In fact, the natural environment has a tendency to treat water itself. As cities grew and urban density increased, however, water quality deteriorated.

The reality is that humans tend to pollute their surrounding environment, including water bodies. Early sources of pollution were mostly human excrements (our pee and poop) and waste related to human activities (from food waste to the disposal of animal waste). Since the industrial revolution, humans have also created new sources of water pollution, including industrial outputs and fertilizer overuse, among many others. As long as human presence is small, pollution remains small. Natural processes such as dilution and biodegradation are enough to treat water naturally. As human presence increases, however, and as new sources of pollution emerge, water bodies start to get contaminated to such a point that natural processes are not sufficient. Consider a series of towns along a single river (which is the case for virtually every large river on Earth). The towns upstream may be able to drink and use water in its near-raw form, but if they empty their sewers in the river, the water available for cities downstream will probably be contaminated and unsafe to drink. This type of situation is common throughout the world. It even caused a big dispute between Chicago and St. Louis in the early 1900s that went all the way to the U.S. Supreme Court. Chicago famously reversed the flow of its main river by building canals and dredging the bottom of the river (to change the slope direction) so that its wastewater would flow toward the Mississippi River instead of Lake Michigan (its drinking water source). But all cities downstream of Chicago on the Mississippi River, like St. Louis, were not happy since they used the river as their main drinking water source.

Ideally, the solution is not to contaminate water bodies in the first place. In fact, many cities have access to clean sources of water that require minimal to no treatment—for example, springs and unpolluted deep wells.[1] But sources of clean water are usually limited, and, as we just saw, when human presence increases, pollution increases. So how can we make sure the water is safe to drink?

Initially, turning to water treatment was not obvious. The main reason we now treat water is so people do not get sick, but for the longest time, we did not know that people could get sick by drinking untreated water. In fact, people once thought that disease spread through "bad" air. That was called the miasma theory.[2] Research in the nineteenth century showed that it is not bad air that spreads diseases but germs—the germ theory of disease—and those germs can be present anywhere, including in water. Famously, Dr. John Snow stopped a cholera epidemic in London in 1854 by having the handle of a contaminated well removed so people could not get water from the well anymore. It was then a matter of time before efforts began to treat water to make it safe to drink.

Generally, the preferred method of water treatment is to take raw fresh water and remove the bad stuff from it through a series of processes described later. Most cities around the world follow this practice. With

---

1. Most of Switzerland is supplied with spring water that receives little to no treatment. See the story at the end of the chapter to learn about Bern. Water from deep wells (technically called *confined aquifers*, as seen in the previous chapter) can be not only safe to drink but also tasty! The Berkeley Springs International Water Tasting competition that occurs every year in Berkeley Springs, West Virginia, often finds that the "tastiest" water comes from municipalities that get their water from deep wells.
2. Some people thought we could even gain weight by smelling food.

climate change and urbanization, however, access to fresh water is not always possible. Desalinating seawater has become a popular alternative. It entails removing everything from the water except the water molecules themselves. Another strategy is to make pure drinkable water, for example, by distilling water through boiling or by reacting hydrogen and oxygen.[3] These processes require a lot of energy, however, and no viable, large-scale technologies have been developed to date.

When it comes to water treatment, few countries are as advanced as Israel. Israel is a relatively small country located in the Middle East—an area of the world that does not get much water. Water is therefore a fundamental resource that is carefully monitored and controlled primarily by one national utility company: the Mekorot National Water Company (in contrast to most cities/regions in the world that have their own, independent water utility company, whether publicly or privately operated). Because water is such a scarce resource in Israel, as of this writing, the average per capita consumption of water is low for a high-income country, with 140 liters per day[4] (compared to 310 liters in the United States, for example).

Despite amazing conservation efforts by the Israeli population to lower their need for water, the availability of fresh water is so limited that Israel has become a leader in desalinating seawater. In fact, over 80 percent of all the domestic water consumed in Israel was originally seawater, and this proportion is expected to exceed 90 percent. Israel's reliance on desalination is not unique in the world (and is increasingly becoming common in the face of climate change). Therefore, we will discuss both

---

3. Although this process needs a spark, and we all know about the Hindenburg disaster (if you do not, look it up).

4. This value accounts only for the water consumed at home. If we include all domestic and public uses (e.g., for commercial and small industrial use like water consumed in offices, shops, and schools), per capita water consumption increases to 240 liters per day, which is still significantly low compared to 520 liters per day in the United States as of this writing. If we want to be completely inclusive, we also need to account for the water consumed for other uses such as agriculture and to generate electricity, but that water is often reused. For example, the wastewater generated by humans is treated and used directly for agriculture purposes in Israel. The values reported here were collected at different times in the 2010s.

the traditional way to treat fresh water and a more recent method to desalinate seawater.

The title of this chapter specifically mentions Tel Aviv, but the water treatment practices that we will see are not specific to the city. In fact, of the two treatment plants that we will visit, one is not even located in Tel Aviv. The country of Israel has a massive water transmission system—the national water carrier—that in part brings water from the north of the country, which is slightly richer in water, to the water-poor south through massive pipes. Having a national water transmission system is not common but essential for some water-poor countries.[5] Operating a national water transmission system is quite like transmitting and distributing electricity, which we will see when we travel to Cape Town and Chicago.

I imagine Tel Aviv as a vibrant and modern city with skyscrapers soaring from the yellow earth alongside historical light-brown buildings that are featured in every movie that takes place around the Mediterranean Sea.[6] I picture myself strolling around busy streets, smelling all sorts of delicious Mediterranean food and spices. A quick online image search of the term *Tel Aviv* also results in a gorgeous beach, a bright blue sky, and a scorching sun that must make one thirsty for delicious and perfectly treated tap water.

In Israel, residents mainly drink from the seas: the Mediterranean Sea on the one hand (through desalination) and the Sea of Galilee on the other, which, despite its name, holds fresh water.[7] Israel also taps into two more freshwater sources: groundwater wells (from two main aquifers) and spring water from the mountains in the north. The technique to treat fresh water is more or less the same throughout the world.

The Sea of Galilee—also called Lake Kinneret—is a large lake located in northern Israel, with an area of 166 square kilometers (64 square miles),

---

5. Especially for smaller countries like Israel located in arid areas. As we saw in the previous chapter, even China has a large water transmission system, which brings water to the water-poor north of the country.

6. Light-colored surfaces reflect rays from the sun to keep buildings cooler in the summer.

7. The Sea of Galilee is the lowest freshwater lake on Earth.

representing an excellent source of water.[8] It is the outlet of a catchment area that covers 2,730 square kilometers (1,050 square miles). One main treatment plant treats the water from the lake and distributes it to a large portion of the country, including Tel Aviv. The plant is aptly called the Central Filtration Plant. Unlike most water treatment facilities, a series of treatment processes are applied before water even reaches the Central Filtration Plant.

From the lake, water is pumped and transported seventeen kilometers through the Beit Netofa Canal to the Eshkol reservoirs, located just upstream of the treatment plant. In the canal, alum and sulfuric acid are added to the water. The main purpose of these chemicals is to favor processes of *coagulation* and *flocculation*. Coagulation uses chemicals[9] to neutralize the electronic charge of suspended particles such as clay—when particles are negatively charged, they repel one another and stay suspended. Once neutralized, the suspended particles can start to aggregate together into clumps called *flocs*; this is flocculation. The newly formed flocs then become heavy enough to settle down or too large to pass through filters (more on those later). This dual process of coagulation and flocculation contributes to lowering the *turbidity* of the water (turbidity is a measure of the cloudiness of the water).

Through the canal, the water is transported to the two Eshkol reservoirs. In the first reservoir (the *sedimentation* reservoir), the water flow slows down, giving time for more flocs to settle (the process is called *sedimentation*); a barge periodically removes the sludge at the bottom, which is dried and sent for burial. The water is then sent to the second reservoir (the *operating* reservoir). Despite having low turbidity, living organisms from the lake (including algae, snails, and other small organisms) are still present and, in fact, have time to grow in the reservoirs. Instead of adding

---

8. In fact, it is such a great source of water that too much water was pumped out of it for consumption, which partly led the push to develop more viable desalination technologies.

9. Besides alum (aluminum sulfate) and sulfuric acid used at the Central Filtration Plant, other chemicals include iron(III) sulfate and iron(III) chloride. Sometimes some polymeric aids are also added to facilitate the flocculation process (although they are more popular for wastewater treatment, as we will learn in Paris).

more chemicals, Mekorot has innovated by using fish that naturally eat the organisms, leveraging ecological principles.[10] Tilapia feed on algae, and the common carp feeds on other organisms. The common carp also churns the mud at the bottom of the reservoir, which both limits the growth of other algae and aerates the bottom. Aerating water increases the amount of dissolved oxygen that can react with certain elements that are dissolved in raw water, like some metals[11] that get oxidized and turn into solids (i.e., they precipitate out and can be removed easily). Over time, other fish species have been added into the reservoir to help treat water naturally.

At this point, the water has a low turbidity and a fairly high quality, but it is not safe to drink yet. It then needs to be transported to the Central Filtration Plant, where it will go through five processes.

First, from the operating reservoir, water flows by gravity through two parallel underground channels, and some chlorine dioxide is added. Adding chlorine or ozone already at this stage is common to disinfect the water and to help remove bad taste and odors. Chlorine and ozone are also effective oxidizing agents (i.e., they precipitate out some elements, like the process of aerating that we just saw).

Second, water goes through a high-rate mixing chamber, where it spends less than ten seconds. Like the process that occurred in the previous channel, the goal here is to add chemicals (the same chemicals, in fact—alum and sulfuric acid) and mix them thoroughly to get a homogenous liquid to favor coagulation. While many particles are removed before water even enters the plant, we now need to be thorough.

Third, water enters a slow-mixing chamber to favor flocculation of the now-neutralized suspended particles. Here, water spends about seven

---

10. A similar process exists in agriculture. Pesticides are used to get rid of insects, but this process has many negative impacts; for example, birds have no insects to eat anymore, or they eat toxic dead insects. Nature is not a series of separate processes but a whole network of processes (aka ecological networks), and changing one of these processes has ramifications on the entire natural system. If you have not read the book *The Silent Spring* by Rachel Carson, I highly recommend it. Personally, it changed my life and how I see nature.

11. Although, usually, the goal is not to remove all metals—some, like iron and manganese, are good for us—but to lower their concentration.

minutes. This is enough time for flocs to form and settle at the bottom of the chambers.[12] At this point, water is clear, but, again, it is not safe to drink.

Fourth, we have the actual filtration process. Chambers are filled with two-meter-deep anthracite (hard coal) granular beds—sand is also used in water treatment plants. Water trickles through the granular beds at a rate of twenty meters per hour. Essentially, this process filters out most particles larger than one or two micromillimeters (including any remaining flocs). After that, what is left in the water is mostly dissolved minerals that are essential for human health. Any bacteria or viruses remaining will be treated with a disinfectant added later. Because the anthracite gets clogged by larger particles that get trapped, air is pushed periodically from the bottom to aerate and clean the filtration beds. The water collected at the top is removed and put back in the operating reservoir to go through the process again later. This last process is called *backwashing*.

Overall, the entire water treatment process at the Central Filtration Plant takes eighteen hours.

Finally, water is pumped up to a reservoir where it will enter the water transmission line that will transport it to various cities around the country. To ensure that the water does not get contaminated in the process, a disinfectant, like chlorine, is added.[13]

This is it. Water is now safe to drink. To ensure that the system works properly, sensors are placed at strategic locations. Separate laboratory experiments are also carried out to measure the water quality.

Around the world, various processes can be added or removed depending on the quality of the collected water. If heavy metals or other contaminants are present in the water, certain chemicals are mixed with the water to create heavy and large flocs using the same coagulation and flocculation processes discussed above. If water is highly contaminated

---

12. This coagulation-flocculation process is sometimes not needed in water treatment plants that collect groundwater since the turbidity is lower (i.e., it has few to no suspended particles).

13. At the Central Filtration Plant, chlorine, ammonia, and soda caustic are added. This process is important since biofilms form in the pipes from any remaining undesirable elements, and adding these chemicals can help prevent contamination of the system.

with microbes, ozone or ultraviolet (UV) light is sometimes used to kill them. As we all know, boiling water kills microbes as well, which would make it safe to drink; however, boiling at this scale requires an excessive and unnecessary amount of energy. Usually, tailored processes are implemented to target specific contaminants present in the raw water. As a result, water quality tests are performed frequently to determine whether new processes need to be added. This is also why no water treatment plants are the same. They are all tailored to the source of raw water that they collect.

The story is different for seawater. In fresh water, we want dissolved minerals to be present because they are essential for human health. In seawater, excessive amounts of dissolved salts are undesirable for human health. Many strategies exist to turn seawater into potable water. From solar and vacuum distillation to multistage flash distillation, turning seawater into drinking water is perhaps the water equivalent of turning lead into gold. Or maybe not, considering that turning seawater into drinking water is actually possible. Over the years, many techniques have been developed. One has revealed itself as more viable than others. To me, its name sounds like the name of a Greek or Egyptian goddess: reverse osmosis.

To learn about reverse osmosis, we travel south from Lake Kinneret and stop fifteen kilometers south of Tel Aviv at the Sorek desalination plant. The Sorek plant is one of the largest desalination plants in the world. It can treat about a third of the water volume that the Central Filtration Plant can treat. It started operating in 2013 and is a marvel of engineering. What is perhaps even more impressive is that the whole treatment process takes only about forty minutes, and it all starts in the sea.

First, water is channeled by gravity from around one kilometer into the sea into two large concrete conduits[14] that bring raw seawater around

---

14. The goal is to channel seawater at a certain rate. To achieve this rate, you can use a small pipe and force a high velocity by having a steep gradient (like we saw in the previous chapter) or use a large pipe and have a slow velocity through a small gradient. Because water comes from the ocean

2.5 kilometers inland where the desalination plant is—an entire trip of 3.5 kilometers. At this point, the water is full of whatever you find in the sea (including plenty of fish) that is screened out.

Second, water goes through an initial pretreatment process to remove as many particles as possible. The process is not unlike the one from the Central Filtration Plant. It includes mixing, coagulation, flocculation, filtration, and backwashing. We will not repeat the steps here. We should note, however, that filtration is taken seriously and micronic filters are used after conventional anthracite and/or sand filters to trap almost all particles larger than one or two microns in size, since we want water to be as clear as possible before getting to the reverse osmosis process.

At this point, the pretreated water is relatively clear, but it still has salt and other minerals that are even finer than what the filters can trap. It is time to move to the reverse osmosis treatment process that is at the heart of desalination.

Conceptually, the treatment process is simple: A polymer membrane made of three layers is manufactured to have a pore size of about 0.0001 micron,[15] which is so small that only water molecules can pass through. Everything else is too big and has to stay on the other side. It is the opposite of the *osmosis* process.

To understand how *osmosis* works, take a permeable membrane and put two liquids around it. Naturally, the two liquids will mix to make a homogenous liquid. Reverse osmosis is therefore the opposite: we start with a homogenous liquid and separate it into two liquids. Because the process is not natural, we need a lot of energy to make it happen in terms of pressure.[16] Essentially, we push the pretreated water at a high pressure

---

and because you do not want to create a large vacuum effect that sucks everything from a large radius, a large pipe and a slow velocity are preferable here. In Sorek, the velocity is kept at 0.15 meters per second, but the pipe diameters are large, somewhere around 4.5 meters.

15. Take one millimeter, divide it into 10 million, and you have 0.0001 micron. If you prefer inches, take one inch, divide it into 4 billion, and you have 0.0001 micron.

16. In thermodynamics, this phenomenon is captured by the second law. Naturally, two liquids at different temperatures will merge into an equilibrium to have the same temperature, unless work is put into it. Here, to artificially separate the two liquids, a lot of work is put into it, and it requires a lot of energy.

on the membrane to force the pure water through the pores. In the end, we get two liquids: pure water that we collect for consumption, as well as brine (i.e., the remaining water heavily concentrated in salt). In some desalination plants, the brine undergoes reverse osmosis again to extract more pure water. The remaining brine is then released out into the sea slowly so as not to disrupt wildlife. At the Sorek desalination plant, another pipe goes close to two kilometers into the sea to release the brine at a depth of twenty meters.

Pressure is therefore key in the reverse osmosis process. At Sorek, this pressure is about 70 bars—picture a 700-meter (2,300-foot) column of water above you. Creating that pressure requires tremendous energy, which is often the main critique about desalination compared to conventional treatment.[17] At Sorek, this energy comes from a natural gas power plant that produces large amounts of electricity to power the hydraulic pumps that create this 70-bar pressure.

The way the membranes are laid out is as follows: A flat membrane is rolled up in a spiral (looking like a galaxy or Cinnabon) and sealed in a tube that is called a *module*. At Sorek, each module has a 40.5-centimeter (16-inch) diameter and contains eight rolled-up membranes coupled in series. This is big compared to many other desalination plants, where modules have only six membranes with a diameter half the size (20.25 centimeters or 8 inches). But that is not even the impressive part. Sorek has 10,000 modules (that is 80,000 membranes), which explains how such a high volume of seawater can be desalinated. In each module, pretreated water is forced on one end, perpendicular to the membrane. As it is pushed, first, pretreated water occupies the entire vertical length of the module. As it is pushed further, pure water enters the pores in the membrane into empty space and trickles to the other end of the module.

The final product is pure water. While it is safe to drink, it lacks essential minerals that humans need that get stuck with the salt in the

---

17. The process can be costly, or at least it is more costly than conventional means. Moreover, it is not environmentally friendly since it requires large amounts of energy, which is provided here by fossil fuels.

brine. A process of remineralization is needed to "harden" the water before it can be sent to the national water carrier. At Sorek, limestone is added to the water to supply it with calcium. Some disinfectants are added as well, similar to what is done with water that comes out of the Central Filtration Plant before it is sent into the national water carrier.

As mentioned above, while being extremely energy intensive, the entire process at the Sorek desalination plant takes only about forty minutes[18] compared to eighteen hours for the Central Filtration Plant, but the comparison is not meaningful. Treatment times do not matter much. Different raw water sources require different treatment processes. Plus, a selected treatment technology may take more or less time. A slower process may also be cheaper and require less energy, which is preferable. In the end, the only thing that matters is that sufficient volumes of water get treated to meet the needs of the population.

As a last note, reverse osmosis does not work solely with seawater. In fact, it could be used to treat water from Lake Kinneret as well, but this is not desirable. First of all, as we saw, reverse osmosis is expensive and energy intensive. Second, treated fresh water has many important minerals that are needed for human health. While desalinated water is remineralized, some important minerals may still be missing. So, if your water comes from a desalination plant, make sure to eat plenty of healthy food that will provide all the minerals that are good for you.

Moving from virtually no water treatment until the end of the nineteenth century, many cities now actively supply treated water[19] to their residents. Now that we know the two techniques to treat water—the more conventional technique to treat fresh water and the more recent technique to desalinate seawater—the next question is: How does the treated water get to the people that need it? To learn about water distribution, we travel from Tel Aviv to the Far East and stop in Hong Kong.

---

18. Although I am not sure this time includes the entire series of processes.

19. Although *treated* does not mean *drinkable*. Check with your water utility to confirm whether the water that flows out of your tap is safe to drink.

THE FOUNTAINS OF BERN

Bern is the capital of Switzerland. It is located roughly in the geographic center of the country, balancing the interests of the German-, French-, Italian-, and Romansh-speaking populations. Selecting a capital city for its location is not uncommon—that is how Washington, DC, in the United States and Ottawa in Canada became capitals. Besides being a capital, Bern is also famous for its fountains. The city boasts over one hundred fountains, some of which have been supplying water for centuries. Each fountain is unique. Many feature a fantastical figure who seems to come out of a Swiss clock. In the city center, one fountain in particular attracts crowds. It is called the Kindlifresserbrunnen, and it has been quenching the thirst of the Bernese since the 1540s. The fountain features an ogre eating a child with its right hand while holding on to several children with its left arm. The whole scene appears to be from a Brothers Grimm story or a Bosch painting. The statue is colorful. The scene is austere but nonthreatening at the same time—at least to adults. I am sure kids must find it terrifying.

Not only did the research for this story make me want to visit Bern and see its fountains, but it especially made me want to drink water from these fountains. Bern is supposed to have some of the cleanest water on Earth.

What is the easiest and most effective way to get water that is safe to drink? It is to collect water that is already safe to drink in the first place. Switzerland is sandwiched between the Alps (south and east) and the Jura (north and west) that both provide abundant, clean fresh water. In the country, the three main freshwater sources are spring water, groundwater, and lake water.

When I hear *spring water*, I imagine a small stream of water—too small to provide for an entire city. But we must remember from the water collection chapter on Rome that soil is porous. Although a water flow may appear small, the soil beneath the spring may be full of water that flows down the hill underground. In Switzerland, spring water is collected by building reservoirs into the sides of mountains. This way, water naturally follows the slope of the land and flows into the water treatment plant. The term *treatment* is a little stretched here. The common practice is to have a chamber to let sediments settle to the bottom. Clean water overflows from this first chamber into a second chamber that has an intake pipe fitted with a sieve to prevent any particle from entering the urban water distribution system. The water quality is so high that chlorine or another type of disinfection process is not needed. That's it. When it comes to treatment processes, it does not come simpler than that.

For groundwater, wells are drilled and pumps installed in sites where the water quality is already high and there is plenty of water. The main concern is to ensure that the wells are drilled upstream of any major agricultural activity that uses fertilizers (overloading the groundwater with nitrogen and phosphorus) or toxic substances like pesticides or herbicides. Once the water is pumped to the treatment plant, the process is identical to that adopted for spring water. Namely, there are two chambers: one for sedimentation and one with an intake pipe (with a sieve) to send the water into the distribution system.

For lake water, the treatment process is slightly more involved, but the water quality from the lake still tends to be high already. Lake water is even supposed to be readily drinkable before any process is applied to it. The few sources I found do not even mention coagulation/flocculation in Switzerland. The water treatment process consists only of filtration using sand. The Swiss are also not fond of chlorine and try to avoid using it.

The water in Bern is supplied mostly from groundwater and a little from spring water. Easy-peasy. It must be boring to manage a water treatment plant in Bern. There is barely anything to do.

I don't know about you, but writing this story has made me thirsty. It is time for a glass of water before we move on to learn about water distribution.

CHAPTER THREE

Water Distribution

Quenching the Soaring Dragon in Hong Kong

HONG KONG. HOW CAN ONE GO ABOUT DEFINING HONG KONG? THE official name of Hong Kong is Hong Kong Special Administrative Region of the People's Republic of China (HKSAR). Hong Kong is so special that it even has the word "special" in its name.

When I wrote these lines, I had been in Hong Kong three months earlier, for the second time in my life. I loved the place as much as I did the first time. I had taken the bus to reach my hotel from the airport, and on the way, Hong Kong appeared to me as a dragon soaring from the turquoise-blue sea. I could see all its natural and lushly green mountains as well as its abundant and colorful artificial mountains (i.e., the endless rows of skyscrapers), starting from the sea and following the topology of the land. The journey was stunning. The landscape seemed surreal even though I had seen it before. In about an hour, we had crossed three massive bridges, traveled through elevated expressways, gone through a large tunnel to cross Victoria Harbor, seen adorable Hong Kong tramways, and even got a taste for the craziness of the crowds in the busy streets of Hong Kong Island. Hong Kong has amazing infrastructure, but it has not always been this way. Joseph Kessel's book *Hong Kong and Macao* details all the extravagances and miseries of the people of Hong Kong in the 1950s.[1]

---

1. Joseph Kessel tells the story of his trip to Hong Kong and Macao in his travel diary of the same name published in 1957.

Those days are now long gone. Hong Kong is not what it used to be. The opium days are over. Kowloon Walled City does not exist anymore. Hong Kong has become a place of wealth and an international financial center. But all of this development relies on one infrastructure system that is nearly invisible to the eye, without which the region could not survive.

In a place like Hong Kong, where elevation (or difference in elevation) rules the lay of the land, any water distribution system must be carefully engineered to provide water to people. Once water has been collected and treated, it needs to be distributed so that when you open your tap, water flows out. Around the world, millions of kilometers of pipes are laid in streets to carry water to millions of buildings. Running this intricate and complex network of pipes and making sure that there is sufficient pressure for water to go where we want it to go can be very tricky.[2]

The first thing to know about water in Hong Kong, even before learning about water distribution, is that water is scarce in this region. Hong Kong does not have large natural lakes, rivers, or significant groundwater aquifers. As a result, since the 1960s, Hong Kong has been importing water from Dongjiang (the East River) in China, representing as of this writing 70–80 percent of all its water consumption. The rest is collected locally, from precipitation—that is, rainwater is collected before it gets to the sea as it naturally would when it follows the water cycle.

In 2023/2024, over one billion cubic meters of water was consumed in Hong Kong. While the number may appear large, it translates to a direct domestic consumption of about 150 liters per day per person.[3] Part of the trick for such a relatively low level of water consumption is that Hong Kong has a second water distribution system. Beyond the freshwater

---

2. Most statistics in this chapter are for the year 2023/2024 and come from the Hong Kong Water Supplies Department, including from the page https://www.wsd.gov.hk/en/publications-and -statistics/pr-publications/the-facts/index.html and the PDF document https://www.wsd.gov.hk /filemanager/common/pdf/Statistics/Keyfacts_2023.pdf (accessed August 20, 2024).

3. *Direct domestic* means consumed by the person at home only. We already discussed this concept in the last chapter: in Tel Aviv, the value is 140 liters per day per person; in the United States, it is 310 liters per day per person. If we include all uses, then water consumption in Hong Kong is closer to 400 liters per day per person, which is still better than 520 in the United States but pales in comparison to 240 liters per day per person in Tel Aviv.

distribution system, it has a seawater distribution system that supplies mildly filtered water mainly used for flushing and firefighting. After all, flushing toilets represents around 20–30 percent of our water consumption, and most cities in high-income countries use drinkable water for toilet flushing.[4] Firefighting also does not require drinkable water. Since the 1950s, Hong Kong had decided to use seawater instead and build a parallel distribution system that covers about 85 percent of the service area. More recent initiatives in the world now recommend the use of gray water. Gray water is water that has already been used but is still usable, like sink and shower water (i.e., virtually all water except for toilet water). The use of seawater is clever, especially since not all buildings could have been connected to the freshwater distribution system in the 1950s, while having a running toilet is essential to improve sanitation. As a result, on top of the one billion cubic meters of water consumed annually in Hong Kong, an additional three hundred million cubic meters is consumed from the sea.

Let us get back to our first water distribution system, the one that distributes fresh water. Essentially, the goal is to take water from somewhere, like from water treatment plants, and push it through a network of pipes across a city so that it reaches every single building. By 2023, Hong Kong had close to 7,000 kilometers of freshwater pipes and 1,700 kilometers of seawater pipes laid in the streets. The material used for seawater pipes is also different; these consist of plastic pipes (PE and PVC) to minimize the effect of corrosion, compared to the ductile iron for mains that transport fresh water. In the end, maintaining water pipe networks represents a monumental effort.

To beat gravity, pressure must be given to water, and there are two ways to get this pressure. One is to have the origin at a higher elevation

---

4. But every solution has its problems, too. There are at least two issues with using seawater. The first is that the volume of water used cannot be metered—the salt clogs meters—but that is not a significant problem. The second, more important concern is the fact that, in Hong Kong at least, both seawater and used drinkable water end up in the same sanitary sewer, which means that the treated wastewater cannot be reused. This situation is not ideal in the age of circularity (which we will learn more about in the next chapter and in the chapter on infrastructure in Singapore), but then we need to remember that 20–30 percent of fresh water is saved.

than the destination. That is what the Romans utilized. The second solution is to use pumps that transfer their mechanical energy to pressure and velocity. That is what most cities do nowadays. Technically, pressure is referred to as *hydraulic energy* and velocity is referred to as *kinetic energy*.

In water distribution systems, the goal is to apply enough hydraulic energy to water so that it reaches a destination (e.g., a building) with sufficient pressure and so that when the tap is opened, water flows out. Or at least, this is the secondary goal. When it comes to regulating water pressure, the primary goal is usually to have sufficient pressure in the fire hydrants to put out a fire. In other words, during a fire, you should be able to connect a fire hose and have sufficient pressure to be able to put the fire out. So what pressure is considered sufficient? Let us talk units.

You may have learned in school that pressure is expressed in pascals (Pa), pounds per square inch (psi), bars, or atmospheres (atm). In water, depending on who you talk to, we prefer to talk in meters (m), pascals (Pa), or pounds per square inch (psi). The three units are linked.[5] Here, we will use meters because it is easier to visualize—picture how high the water goes when you point a fire hose toward the sky. To determine what pressure is sufficient, we must determine how high we want our water to reach.

Historically, in many countries, buildings had a maximum of five or six floors—that was before the advent of the elevator. During a fire, you want the water pressure to be high enough to put out a fire on the top floor. That is the basic rule that dictates water pressure regulations. And, by definition, this pressure will also be high enough to reach top floors when you open a tap. This five-or-six-story rule translates into an average pressure somewhere between twenty-five and thirty meters (eighty to one hundred feet). This is why water towers tend to be around forty to fifty

---

5. Pascals (Pa) and pounds per square inch (psi) express the same thing, but Pa is in metric (1 Pa = 1 kg/[m×s²]) and psi is in imperial units. They express pressure directly. To turn them into meters (m), we use the formula $P = \gamma \cdot h$, where P is pressure, γ is the specific gravity of water (density [ρ] × gravitational constant [g]) usually taken as 9.79 kN/m³ for water at 4°C), and h is height in meters (m). Picture a column of water or a well; to calculate the pressure of the water at the bottom of the column/well, we use the equation $P = \gamma \cdot h$.

meters high (an extra ten to twenty meters to account for friction and minor losses that we will learn about later).

In addition, we must account for the fact that water pressure fluctuates across a system. For example, pressure is higher in houses located in a valley than houses located on a hill. Moreover, when people consume a lot of water at the same time (e.g., when everyone takes their shower in the morning), the average pressure can drop. Another regulation must be in place so that water pressure is never below a certain threshold anywhere in the system at any time. Throughout the world, this minimum pressure is around ten meters, generally for fire safety. Plus, fire trucks now have pumps to ensure sufficient pressure, but the water needs to be there in the first place—hence the minimum pressure requirements.

In Hong Kong, because elevation changes significantly, the regulation argues for a minimum pressure between fifteen and thirty meters (quite a big gap, but, again, elevation rules the lay of the land in Hong Kong). As a result, the average pressure in the region is between sixty and eighty meters. This number is high but unavoidable to make sure the pressure at any point in the system does not drop below the minimum threshold. And how do you provide that pressure in a place that is anything but flat? You make elevation your friend.

As we saw before, in Hong Kong, water is collected from the rain and imported from the Dongjiang. This water must first be stored somewhere (storing water locally ensures that Hong Kong can be self-sufficient for several days/weeks if needed). Usually, raw water (i.e., not treated yet) is stored in open-air reservoirs that look like large ponds or small lakes. They are similar to the distribution basins from the chapter on water collection in Rome. Hong Kong has seventeen of these reservoirs. They are called *impounding reservoirs*. As Hong Kong is hilly, most impounding reservoirs are small. To increase capacity, two giant reservoirs were built into the sea: the Plover Cove Reservoir in 1968 and the High Island Reservoir in 1978. In both cases, dams were built to retain raw water and seal off the reservoirs from the sea.

From an impounding reservoir, raw water is then pumped to one of the twenty water treatment plants located across the region. This is where water is treated using processes like the ones we have seen in the previous chapter. Once treated, water is stored in *service reservoirs*. Service reservoirs are much smaller and enclosed so treated water does not get contaminated. Hong Kong has 170 of them.

The service reservoirs are strategically located. Remember that, ultimately, water must have a pressure between sixty and eighty meters. That pressure can come either from pumps or from gravity itself. Because of Hong Kong's significant elevation changes, gravity can do the work. Essentially, if the reservoirs are placed high enough—above buildings—gravity will ensure that there is always enough pressure in the pipes. As a result, most of these reservoirs in Hong Kong are located on very high ground, close to mountaintops. The reservoirs are also turned into parks, as is common in many cities. Buildings cannot be built on top of reservoirs, but parks can be landscaped, turning service reservoir areas into multifunctional social amenities.

While Hong Kong has many unique features, building service reservoirs at higher elevations is common in places that offer such a landscape. It is the case of many cities in Switzerland. It is also the case of my hometown of Saint Pierre and Miquelon. When elevation is not available, however, pumps need to be used, and it can become a bit tricky. Remember that pressure changes with consumption rates—if a lot of people have their taps open, the pressure will drop. Usually, large cities prefer to avoid pressure changes since they cause pipes to wear out quickly, leading to breaks. In these instances, pressure regulators or pumps that can vary their pumping rates tend to be used—still, all options have pros and cons.

Here, we have only talked about gravity as the single force that water has to overcome to get to a destination, but there are two other forces that must be considered: friction losses and minor losses. Minor losses occur at pipe junctions or when pipes change direction (e.g., pipe elbows).[6] Friction losses occur naturally on pipe walls because of shear forces (i.e., as a

---

6. The equation for minor losses H_M is $H_M = K \cdot \dfrac{V^2}{2g}$, where K is a factor that accounts for the type of loss (e.g., pipe junction versus elbow), V is the flow velocity, and g is the gravitational constant.

resistance to the flow).[7] Both losses incur a loss of hydraulic energy (i.e., decrease in pressure)—technically, the energy is transformed into heat. To make up for it, additional pressure needs to be given. That is why water towers tend to be around fifty meters tall when the regulated average pressure is around twenty-five to thirty meters.

Another factor to consider is flow velocity (i.e., how fast water flows in the pipes). Flow velocity may not be important to consumers (as long as water flows out when a tap is opened), but it impacts minor and friction losses. Essentially, higher velocity translates to higher minor and friction losses,[8] thus into more acute decreases in pressure. Controlling flow velocity therefore helps control the pressure as well. And by minimizing the input energy provided by the pumps, we are also saving electricity. In addition, higher velocity accelerates pipe wear, leading to undesirable pipe leaks and breaks. In the end, flow velocity is not recommended to be above 1.5 meters per second.

Another important consideration in water distribution is pipe leaks. Simply put, they are inevitable in large water distribution systems. Pipes can be in the ground for decades, up to a century or even more, and they are bound to leak at some point. Large leaks (e.g., water main breaks) are easily detectable and are solved by repairing or replacing a pipe. Smaller leaks are not easily detectable, however. Leaks are also called *line losses*. Technically, they make up the difference between *water withdrawal* and *water consumption*. On average, in the United States, between 6 and 16 percent of all water distributed in pipes is lost through leaks.[9] This number is quite low. In 2011, the leakage rate in London was estimated to be over 25 percent, and it was estimated to be over 50 percent in Rio

---

7. A typical equation used to estimate friction losses H_F is the Darcy-Weisbach equation: $H_F = f \cdot \dfrac{L}{D} \cdot \dfrac{V^2}{2g}$, where f is a friction factor (which depends on the pipe roughness and the Reynolds number), L is the pipe length, D is the pipe diameter, V is the flow velocity, and g is the gravitational constant. From the equation, we see that friction losses are higher in longer pipes with smaller diameters.

8. See the equations in the two previous footnotes.

9. C. M. Chini and A. S. Stillwell, "The State of U.S. Urban Water: Data and the Energy-Water Nexus," *Water Resources Research* 54, no. 3 (2018): 1796–1811.

de Janeiro (Brazil) and Manila (Philippines).[10] In Hong Kong, the leakage rate was about 14 percent in 2022, down from over 25 percent in 2000.[11] Even in Singapore, which possibly has the most well-maintained water distribution system on Earth, the leakage rate is around 8 percent.[12]

To control the amount of leakage, as well as pressures and pipe velocities, water distribution systems are often equipped with supervisory control and data acquisition (SCADA) systems. SCADA systems leverage telecommunication infrastructure to monitor in real time how a water distribution system is performing. They are integral to large water distribution systems, and they are commonly used in electricity, natural gas, and other infrastructure systems as well.

No matter how well maintained and monitored a water distribution system is, leaks and pipe breaks are inevitable, or at least for as long as we will have water distribution systems that consist of thousands of kilometers of highly pressurized pipes. Water distribution systems are like pressure cookers that are constantly turned on. Maintaining that pressure comes at a cost. Water utilities are often the largest consumers of electricity in cities. On average, the pumps needed for water distribution consume around 80 percent of all the energy used by water utilities (the remaining energy is consumed for water treatment[13]). Water distribution systems are therefore quite vulnerable (i.e., prone to pipe breaks), and they consume a lot of energy. Modern water distribution systems have essentially followed the same model since the end of the eighteenth century, and it might be time to revisit the current strategy. In the following box, we will learn how water is distributed in Hanoi (Vietnam), which might offer us some elements of an alternative.

---

10. C. Kennedy et al., "Energy and Material Flows of Megacities," *Proceedings of the National Academy of Sciences of the United States of America* 112, no. 19 (2015): 5985–90.

11. From https://www.wsd.gov.hk/en/core-businesses/operation-and-maintenance-of-waterworks/reliable-distribution-network/index.html (accessed August 20, 2024).

12. M. Arora, L. W. Yeow, L. Cheah, and S. Derrible, "Assessing Water Circularity in Cities: Methodological Framework with a Case Study," *Resources, Conservation & Recycling* 178 (2022): 106042.

13. Unless water is desalinated, in which case water treatment also becomes energy intensive.

Now it is time to learn about where the water goes after it has been used and sent down the drain: it is time to learn about sewers. And for that, we will leave beautiful Hong Kong and travel west, back to Europe, and make a stop in the French capital: Paris.

DISTRIBUTING WATER IN HANOI

Adversity often becomes a source of creativity. The residents of Hanoi, the capital of Vietnam, are resilient and clever. If you go to Hanoi today, no matter the time, when you open a tap, water will flow out. But this was not always the case . . . or was it? After the Vietnam War (or the American War, as they call it in Vietnam), Hanoi grew and modernized. For some time, enough water was not always available for everyone to use. In turns, certain neighborhoods were periodically shut off from the system and did not receive water (we will see this practice apply beyond water distribution—wait until we travel to Cape Town to learn about electricity).

As water can be stored easily, the people of Hanoi installed water tanks in their basements and filled them up when water was available. This way, even if their access to water was shut off, they could be self-sufficient for several days. Basements were selected for these basins because this is where water arrives. Plus, remember what we learned about pressure: if water pressure is not high enough, it cannot reach higher stories. As water demand was high, pressures were low. Basements simply made sense given that water could easily flow into the tank, even at low pressures. In addition, water is heavy. Having large tanks on higher floors can require stronger (and more expensive) structural elements.

Yet having water stored on higher floors also makes sense. As water follows gravity, having a water tank on the roof ensures that water remains available even during power outages. Many residents installed a second, smaller tank on their roof that would be supplied from their basement tank with a pump. With Hanoi's tropical weather, it also meant that the water could warm up naturally (lowering the need for water heating). The main drawback is that water quality decreases when it stagnates, especially at warmer temperatures, but the water quality from the raw water coming into the basement tank was not high enough anyway (low pressures zones mean contaminated groundwater can enter the water distribution system).

Nowadays, all buildings are continuously supplied with water, but the practice of having a basement tank and a roof tank remains. This situation means that the water pressure does not need to be as high, which in turn means fewer leaks, lower friction losses, and less energy used for pumping. When I lived in Hanoi for

---

about six months in 2019, I was told that the water pressure in the water distribution system was essentially dictated by the fire safety minimum.

While I only drank bottled water, I learned that many Hanoians had installed filters on their taps to make the water drinkable. This practice is generally called point-of-use (POU) water treatment. While the water coming out of the water treatment plant is drinkable, it can still get contaminated in the distribution system—hence the need for POU treatment. Overall, I was stunned. The people of Hanoi had managed to build a water distribution system that required less energy and was more resilient than the systems I was used to back home. I later learned that many other cities have similar systems, including cities in India.

I ended up writing a journal article on the Hanoi water distribution model, criticizing the Western model in comparison. The one comment I received was one of equity. Basement and rooftop tanks are costly. Imposing these costs on low-income families would be unreasonable in higher-income countries. While I agree, I also like to point out that Vietnam and India are not rich countries, but they have made it work.

What I particularly like about this story is that innovation came not from the water utility company but from the consumers. In higher-income countries, infrastructure services are provided to us without us needing to understand anything about them. Maybe it is time to revisit this model. Maybe we are taking some of these services for granted. Maybe consumers should be more responsible for the services they consume.

In the end, my only certainty is that the current water distribution model must evolve, and Hanoi has shown us a different model from which we can learn.

Chapter Four

Sanitary Sewers

Conscience, Convergence, and Clarifiers in Paris

Let the reader imagine Paris lifted off like a cover, the subterranean network of sewers, from a bird's-eye view, will outline on the banks a species of large branch grafted on the river. On the right bank, the belt sewer will form the trunk of this branch, the secondary ducts will form the branches, and those without exit the twigs.

. . .

The sewer is the conscience of the city. Everything there converges and confronts everything else.

—VICTOR HUGO, *LES MISÉRABLES* (1887)[1]

PARIS, THE CITY OF LIGHTS AND THE CITY OF LOVE.[2] HOME OF SOME OF the most iconic monuments ever built: the Eiffel Tower, the Louvre, the Arc de Triomphe. Not unlike Rome, Paris is an open-air museum. Just

---

1. Volume 5, book 2, chapter 2.
2. Yes, Paris has two nicknames. Paris is the City of Lights because it was the epicenter of the Age of Enlightenment in the eighteenth century when great thinkers came together to share their progressive ideas and also because it is one of the first cities to adopt gaslighting to light its streets at night. Paris is also the City of Love because of its charming architecture and the significant presence of artists and writers in the nineteenth and twentieth centuries that gave Paris the romantic image we all know.

wandering around the city offers a unique experience.[3] One of my favorite spots in Paris is the Pont des Arts, a pedestrian bridge built in 1804, located at the nexus of some of the most beautiful views of Paris: the Louvre to the north, the Île de la Cité to the east (with a great view of the Square du Vert-Galant), the Institute of France to the south, and the Eiffel Tower to the west—I spent the COVID-19 lockdown in spring 2020 in Paris, and the Pont des Arts was within the radius I was allowed to visit (which I did, often).[4] I also wrote many chapters of this book in Paris during my stay.

In this chapter, we will discover another Paris. Although less charming, the Paris that we will learn about remains fascinating nonetheless. By the end of this chapter, Paris may be given yet another nickname: the City of Sewers.[5]

As the idiom goes, what goes in must come out. That notion particularly applies to water. So far, we have put a lot of effort into collecting, treating, and distributing water. We now need to figure out what to do with it once it is consumed—that is, once water turns into sewage. Dealing with sewage was not a problem for the longest time, simply because people did not consume much water. By the 1700s, people in Paris consumed about five liters of water per person per day[6]—that is a bit more than one gallon for drinking, washing, cooking, and so on, for an entire day. It is not that people did not need water; rather, it is that water was not easily available. With a consumption level of five liters per day, extensive infrastructure to deal with sewage was not needed.

---

3. In the book *Connaissez-vous Paris?* (*Do You Know Paris?*), Raymond Queneau describes the activity of wandering around a city as an "antiopée," which has been one of my favorite activities for a long time, not only in Paris but in cities in general.

4. Except for work or for essential activities (like buying groceries), people living in France were allowed to walk only within a one-kilometer radius of their home (a bit more than half a mile). Since I lived in the center of Paris, I got to see the surroundings of the Louvre, the Rue de Rivoli, the Pont des Arts, and the Halles, all virtually empty. I feel privileged despite the tragedy of the pandemic.

5. A portion of the Paris-specific content and data for this chapter was collected from SIAAP, "L'assainissement des eaux usées en agglomération parisienne: Principes et procédés," Service public de l'assainissement francilien, 2013.

6. Historical water consumption statistics in Paris can be found in S. Derrible, "An Approach to Designing Sustainable Urban Infrastructure," *MRS Energy and Sustainability* 5, no. 1 (2019): 13.

Over time, water became more available, partly thanks to early distribution systems that could feed water to fountains across cities. As a result, water consumption increased, first to about 10 liters per person per day in the early 1800s. In Paris, by 1850, water consumption was estimated at around 20 liters per person per day. In the latter half of the nineteenth century, as cities grew substantially, and as hygiene became a predominant concern around the world, more water distribution systems were built that made water available to individual buildings. From having to walk to a well, crank up a handle, and fill and transport buckets, people now could stay in the comfort of their homes and get their water by opening a tap. As expected, water consumption shot up, reaching about 100 liters per day per person in the late 1800s. By 2012, a person in France consumed an average of 150 liters per day (not bad compared to Israel's 140 liters per day), and in Paris, this number was closer to 120 liters per person per day. As a bit of trivia, Paris is also home to several free municipal sparkling water fountains around the city. Next time you visit the city, locate one of them and fill up your bottle with sparkling water.

The jump from five to one hundred liters per day by the end of the 1800s is huge. No matter how many times they were emptied, cesspools and other in-home strategies of containing sewage were not enough anymore. While a higher consumption of water was desirable in the first place to promote hygiene and extend life expectancy, the resulting sewage became a nightmare and required the development of an altogether new type of infrastructure: the sanitary sewer.

Before we discuss how sanitary sewers function, we must go back one step. There are two types of wastewater (i.e., water we want to get rid of in a city): sanitary wastewater and stormwater. Sanitary wastewater comes from human activity (e.g., water used in showers, sinks, toilets, and so on). Stormwater is all the water that falls and flows in cities when it rains. Cities have always had to deal with stormwater. As a result, many cities have been building infrastructure to deal with stormwater for a very long time (going back thousands of years). This infrastructure was often in the form of drains, such as ditches on the side of roads, that would fill with stormwater during rain events and drain it away using the same principles as the ones we saw in the water collection chapter (i.e., using gravity as the main force). Roads were also designed to channel stormwater away—roads are arced so that stormwater runs to the gutters on each side to then flow to catch basins. For several centuries, some paved roads (especially in France) were sloped toward the center so that one single central gutter could carry the stormwater away and wash out any litter.[7] Similar to contemporary stormwater sewers, large drains were placed by some cities under the roads to channel stormwater away. We will discuss stormwater management at length in the next chapter, but it is important to remember here that some infrastructure already existed in cities to deal with stormwater.

Now, say you are a city official, and you need to address the sewage problem—that is, the fact that more and more sanitary wastewater is generated and cannot be contained in buildings anymore. What do you do? The main idea is to have some kind of conduit—buried because of the smell—to bring sewage from buildings to somewhere outside of the city, preferably downstream of a river. Downstream is better so as not to pollute the upstream source of fresh water often used for drinking. Downstream is also necessarily at a lower elevation, and again, we can use gravity to transport the sewage, similar to water collection. But then, how do you do it?

One idea is to build a completely new system, and another idea is to connect buildings to underground stormwater drains that already exist. This is the birth of the combined versus separate sewer dilemma. As the name suggests, combined sewer systems combine sanitary and stormwater sewers. They are cheaper to build since there is one single system instead of two, but the system can get full during heavy rains, leading to sewer overflows—technically called CSOs for *combined sewer overflows*. CSOs occur when raw sewage is dumped into natural

---

7. The idea originally came from Philip II of France (better known as Philip Augustus) in the twelfth century. You can still find old streets with this type of design in some cities in France.

bodies of water like lakes and rivers and when sewage backs up in basements. Separate sewer systems, again as the name suggests, have two systems, one for sanitary wastewater and one for stormwater. As a result, there are no CSOs regardless of how much it rains. Also, we have not talked about wastewater treatment yet, but it is important here. Because sanitary wastewater must be treated, in combined sewer systems, all wastewater is treated (sanitary and stormwater), representing enormous volumes of wastewater that need to be treated, significantly increasing the operating costs of wastewater treatment plants. In contrast, in separate sewer systems, only the sanitary wastewater is treated; stormwater is often judged clean enough to be dumped directly into natural bodies of water.

When sewers were starting to get built, which was before sanitary wastewater had to be treated, many cities went with the combined sewer system option—it was so much easier and cheaper—but that is a decision that cities around the world have come to regret. Combined sewer systems are now seen as a bad idea, especially as storm events are increasingly more severe and more frequent with climate change. Indeed, CSOs generate significant negative impacts, both in terms of public health and for the environment, by dumping raw sewage into natural water bodies. In the United States, the Clean Water Act has made CSOs illegal (although they still occur in virtually every city with combined sewer systems). As far as I know, all new cities or new neighborhoods in cities choose to go the separate sewer system route. Some cities with combined sewer systems even work toward converting their combined system into a separate sewer system.

Now, back to Paris and its sanitary sewer system. Originally, back in the nineteenth century, few cities in the world were as proactive at building sanitary sewers as Paris. When he ruled France in the latter half of the nineteenth century, Napoleon III had a plan to transform Paris. He was greatly influenced by London, where he lived in exile for many years. Because of the Great Fire of 1666, London had been profoundly modernized, becoming a model of hygiene for cities around the world. Through

Baron Haussmann,[8] Napoleon III decided to modernize Paris, massively moving populations and destroying entire neighborhoods.[9] By and large, the Paris that we know today dates to that time, not only the city that you see above the ground but also parts of the city that lie beneath the ground—namely, the sewer. One of the chief architects of the Paris sewers was Eugène Belgrand, who started working on the modern system in 1854. Paris did not have a significant stormwater sewer system before then, so a single system was designed and built to accommodate both sanitary wastewater and stormwater. Paris was therefore given a combined sewer system, but it is unlike any other combined sewer system in the world. Later, since the 1970s, following the global trend, separate sewer systems have been built for new cities and neighborhoods in the surrounding Paris region.

In sewer systems, the size of the sewer conduits varies. Small branching conduits are used to connect buildings to sewer conduits in the streets, and these first sets of sewer conduits converge to larger conduits, which themselves connect to even larger conduits to accommodate an increasing flow of wastewater. Sewer systems essentially mirror water distribution systems in which small service lines to individual homes are fed from larger conduits, which are fed from water mains, which sometimes get their water from larger conduits that come out of water treatment plants and service reservoirs. In Paris, the smaller sewer conduits under most streets have a diameter of around one meter (3.3 feet). The larger sewer conduits have a diameter of three to six meters (10–20 feet). These large sewer conduits look a little like the cross-section of a button mushroom. There is a central, deeper channel where the wastewater flows, and the elevation of the water changes depending on the flow of the wastewater (i.e., whether it is raining). On each side, a little higher, two walkways are

---

8. Haussmann was appointed prefect (representative of the national government in the Paris region) and led to the modernization of Paris. The classic Paris architecture that we all know is usually referred to as Haussmann architecture.

9. Although this may sound bad, Napoleon III was attuned to the problems of his people, especially with the deplorable conditions of industrial workers. He is the one who made it legal to go on strike and who established a retirement system.

present that are used for maintenance and for people to visit. Yes, you can visit the sewers of Paris. An entire museum is dedicated to it. It is fascinating and unique in the world. Many writers have also written about the Paris sewers in their novels, including Victor Hugo, who wrote extensively about them in *Les Misérables*, viewing the sewers as the conscience of the city where everything converges (including people).

Like water collection systems, sewer conduits must also be carefully constructed so that they slope in the right direction to use gravity as the main force to transport wastewater. In Paris, because the elevation in the city changes, several pumps have been installed to pump wastewater up a few meters.

But that is not all. Many of the sewer conduits in Paris are also used to house pressurized water pipes—attached to the roof of the tunnels—to distribute water (instead of burying the pipes in the ground, thus helping with maintenance and reducing leakage).[10] Moreover, with the rise of the internet, the sewers are also used to house fiber-optic cables, keeping the physical world connected to the virtual world. The sewers of Paris are therefore much more than regular sewers. They are critical, multifunctional pieces of infrastructure without which the region could not operate properly.

Responsibilities for the different aspects of this intricate network of sewers in Paris are divvied up between two organizations in the region. Every municipality (including the City of Paris for central Paris) is responsible for collecting all its wastewater and transporting it to a regional infrastructure system that is operated by the Greater Paris Sanitation Authority. In total, by 2022, the City of Paris operated about 2,600 kilometers of sewer conduits, including the ones that can be visited. All wastewater is then transported through a 440-kilometer system of large

---

10. Paris actually has two water distribution systems: one for potable water, as we have seen in Hong Kong, and one for nonpotable water that is used for irrigation, to clean the streets, and to flush the sewers (for maintenance). Arguably, nonpotable water could be used for toilet flushing, like the seawater system in Hong Kong, but it would require connecting the nonpotable water system to individual buildings and changing all the plumbing.

tunnels that cannot be visited and that are operated by the Greater Paris Sanitation Authority.[11]

The role of the Greater Paris Sanitation Authority is not only to collect all the wastewater but also to treat it. The region has six wastewater treatment plants and five wastewater pretreatment plants. Treating wastewater can be a complicated affair depending on how clean we want the wastewater to be. In general, the goal is to make it clean enough so that the water coming out does not pollute the body of water in which it is disposed, but not necessarily clean enough to drink right away. In water-scarce regions, however, making drinkable water from wastewater is necessary, and this is when reverse osmosis comes in handy again, since only pure water can permeate the membranes used.

A typical wastewater treatment process involves four steps. Like water treatment, the goal with wastewater treatment is to remove the bad stuff. The first step is called pretreatment. The objective is to remove larger particles using physical means (usually, no chemical reaction is induced yet). With pretreatment, the wastewater goes through a series of screens of varying mesh sizes to remove large particles first, such as tree branches and plastic bottles, and then smaller particles of a few millimeters in size, such as small rocks. The wastewater next goes through a grit chamber to remove finer particles like coffee grounds and sand that can be as small as a fifth of a millimeter.[12] Given some time to settle, heavier particles like residual sand fall to the bottom of the chamber and can be removed. At the same time, because they are lighter than water, greases in wastewater rise and float on the surface, and they can easily be skimmed off. Sometimes air is blown into the wastewater as well to help separate organic and inorganic particles and fat from water molecules. Plus, the increased quantity of dissolved oxygen can help oxidize certain elements that precipitate out, as we saw with water treatment in Tel Aviv. All the waste

---

11. Or the Syndicat Interdépartemental pour l'Assainissement de l'Agglomération Parisienne (SIAAP) in French. The SIAAP also calls its tunnels *wastewater highways* since they operate the large regional wastewater conduits that are not unlike the web of expressways that exist in many American cities, including in Los Angeles, which we will learn about soon.

12. Around 0.2 millimeters or 200 microns, which is about 0.007 inches.

collected that includes particles and grease is usually referred to as *sludge* (we will learn more about sludge later). The pretreatment process is now complete. Wastewater is clearer and void of large particles, but it is not clean. In addition to the six treatment plants, Paris has five pretreatment plants because pretreated wastewater is less likely to damage the large interregional conduits.[13]

After pretreatment comes treatment. Three series of processes are applied to treat wastewater. Like grit chambers, *primary processes* consist of letting finer particles settle and any remaining greases rise to the surface. Primary processes often occur in large circular chambers called *primary clarifiers* that are equipped with large mechanical scrapers that pivot around a central axis to scrape the bottom and skim off the surface to collect the sludge—they are so big that you can see them when flying.[14] An alternative to primary clarifiers is to use *lamella clarifiers* that occupy less space. In lamella clarifiers, inclined tubes are laid out in the chamber (a little like long inclined honeycombs) that help finer particles settle more quickly.

Generally, chemicals are also mixed with the wastewater before clarification to favor the process of *flocculation*, which we first saw in Tel Aviv.[15] Namely, the chemicals help the particles to aggregate into clumps called *flocs* (i.e., this is the flocculation process) that become so heavy that they settle at the bottom of the clarifiers to be scraped off.

In Paris, all of these strategies are used. Processing times differ depending on process; it takes two to five hours for a treatment plant that

---

13. Have you heard the term *fatberg*? It is when grease accumulates in sewer pipes, eventually blocking them. It is not a pretty sight, and it is a big problem. That is why authorities recommend throwing greases in the trash and not in the sink.

14. Wastewater treatment plants are often located near airports. The primary reason is that wastewater treatment plants are built in lower-elevated lands to use gravity for the transport of wastewater. These lower-elevated lands can be swampy or more prone to flooding and thus are less likely to be inhabited and offer available land to build large structures like airports. Next time you fly, make sure to look out the window when taking off and landing.

15. In the Tel Aviv chapter, we learned about *coagulation* and *flocculation*. Coagulation is not as much of a problem in wastewater treatment because the particles are different from those in drinking water. The chemicals used to enhance flocculation can be the same as the ones for coagulation, however. For example, iron(III) chloride is a great coagulant and a great flocculant, notably used in Paris. Another great flocculant is polymeric aids.

uses traditional primary clarifiers, one hour for a treatment plant that uses lamella clarifiers, down to twenty-five minutes for a treatment plant that adds chemicals before the lamella clarifiers.

The goal of *secondary processes* is to deal with dissolved particles, mainly organic, such as sugars, that are by nature dissolved and cannot settle down in a chamber. Dissolved particles can come from various sources, including excrements, food waste, and soap products. Instead of physical and chemical processes, secondary processes leverage biological processes to degrade pollutants and treat water. Put simply, all these dissolved particles are food that macro- and microorganisms like bacteria and fungi ravish. A little like the filtration process in water treatment, wastewater passes through biofilters that can be made of various media (e.g., sand or small plastic beads) and have macro- and microorganisms that eat the dissolved organic particles for their personal growth. In older systems that require more space and time, wastewater is sprinkled on top of biofilters. In newer systems, the wastewater is injected at the bottom of the chamber with some air and goes through the biofilter in an upward direction so that the wastewater at the surface is cleaner. In both old and new systems, the addition of air (i.e., to increase the amount of dissolved oxygen) via sprinkling or injection is also essential to help the biological organisms break down the dissolved particles. Indeed, after all, they are living beings. They need oxygen, too, just like we do. Without oxygen, they cannot live and do their job properly.

As they grow, macro- and microorganisms create a slimy and muddy biofilm around the sand grains or plastic beads used in biofilters, and they must be washed frequently. In Paris, secondary processes also include nitrogen removal (aka *denitrification*) that otherwise would react too much with oxygen[16] once the wastewater is disposed of in natural bodies of water. Denitrification is a two-step process that also leverages different bacteria. First, the ammonia in the wastewater is converted into nitrate that is itself removed in the second step. Some

---

16. Oxygen needed for wildlife like fish.

wastewater treatment plants in Paris even have a third step in the secondary process to remove phosphorus from the wastewater, which, if it occurs in a high concentration, can lead to the eutrophication[17] of water bodies.

This is it for secondary processes. Instead of biofilters, some wastewater treatment plants engineer large surfaces of open wetlands where natural organisms and plants naturally treat the wastewater. Other wastewater treatment plants also include nitrogen and phosphorus removal processes in the tertiary processes.

Tertiary processes sometimes include further filtration and usually end with a final disinfection of the treated wastewater, which can be as simple as adding a little bit of chlorine before releasing the treated wastewater into a natural body of water. In Paris, several tertiary processes are applied, varying from treatment plant to treatment plant, including further phosphorus removal processes and the use of ultraviolet (UV) lighting[18] or a membrane[19] for disinfection.

The time it takes for the entire wastewater treatment process can fluctuate significantly. Depending on the technology adopted, in Paris, processing times vary from fifteen hours down to two hours—although the goal is not to be as fast as possible but as effective as possible.

In the future, a fourth process may be added to address micropollutants such as pharmaceuticals, household chemicals, and pesticides, which can be acute sources of pollution for natural water bodies and are difficult to remove from wastewater.

Moreover, worldwide, wastewater treatment plants increasingly try to capture nitrogen and phosphorus from wastewater to sell as fertilizer—this

---

17. Phosphorus is used in plant fertilizers and therefore favors algae growth in water; algae consume a lot of oxygen as they grow, impacting wildlife. We have all seen examples of eutrophication. That is when a body of water becomes fluorescent green because of an abundant presence of algae.

18. Using UV light is a one-off strategy, unlike chlorine. That is why chlorine or some other disinfectant is preferred for water distribution since it limits the growth of pathogens in the water pipes. For wastewater treatment, this continued disinfection process is not needed, so UV light works well.

19. Although the process is similar to reverse osmosis, the membranes that are used have a larger mesh, and some particles get through in addition to pure water. Unlike drinking water, the use of reverse osmosis is not desirable since it requires a tremendous amount of electricity and is expensive.

step is especially important given that the natural mines of phosphorous are depleting, leaving us on the verge of an agriculture crisis.

There are two more components of wastewater treatment to touch on. The first is odor control, and several processes (including chemical, physiochemical, and biological options[20]) are adopted to rid the foul smell that can be present around wastewater treatment plants. The second is the management of the sludge that is generated during the treatment. There are various things that can be done with the sludge. For example, in a controlled environment,[21] it can be turned into compost and used as fertilizer to grow crops. Or it can be put into an anaerobic digester[22] to produce natural gas that can be used to heat buildings or generate electricity. After processing and drying, the sludge can also be burned in an incinerator to produce electricity and heat. The sludge (or whatever is left, like ash after burning) can then be landfilled. We will learn more about some of these processes when we visit Tokyo to discuss solid waste management. All or most of these sludge management options are used throughout Paris.

Despite being largely hidden from the world in underground conduits, wastewater management is not simple. The collection and management of sanitary sewage have come a long way since the nineteenth century, and Paris offers a wonderful case study that everyone can experience firsthand by visiting the Paris Sewer Museum.

Sanitary sewage is just one of the two types of wastewater. To learn about the other type—stormwater—we will remain in Europe but travel northeast to Denmark. It is time to travel to Copenhagen.

---

20. To transform foul and sometimes toxic compounds into something else that does not smell bad and is not toxic.

21. Controlled because unprocessed sludge can be full of pollutants.

22. Essentially, the sludge is put in an airtight container. As the organisms eat and process the sludge, methane is produced because oxygen is absent. We will learn more about the process in the solid waste management chapter in Tokyo.

THE *POZZO NERO* BOATS OF VENICE

Venice is an extraordinary city, as in, it is beyond the ordinary. No conventional infrastructure solution works in Venice. La Serenissima is unique. Even its origin is extraordinary. Venice is not located on a single island; rather, it is a cluster of islands. While many cities have their origins as a small settlement that expanded, Venice was made of multiple settlements, each of which had many similar features (e.g., a church and a public square [the *campo*]). Even today, you can experience it. There are public squares everywhere. You can also see that even the major islands are made up of small islands, linked by more than four hundred bridges. Look up Venice on a map. It is in a lagoon at the edge of the Adriatic Sea. Twice a day, the water levels go up and down with the tides. That is also why Venice is no stranger to flooding during high tides (*acqua alta*). When it comes to sewers, being so close to the sea is a problem.

If all roads lead to Rome, as the idiom goes, all water ends up in the sea. That is why we like to get our fresh water from places at a higher elevation and use gravity to bring it to a city. Then, in a city, we position the sewers in such a way that gravity does all the work to bring the wastewater to a lower elevation (often downstream of a river), to a wastewater treatment plant. What happens when our city is already at the lowest possible elevation?

In Tokyo, the sanitary wastewater was historically collected and used as fertilizer by farmers. But Tokyo is in an estuary, not a lagoon. Unlike in Venice, the water levels in the canals around Tokyo are not impacted by the tides. In contrast, in Venice, the tides can act as a massive, natural flushing mechanism. The solution selected to deal with sanitary sewage in Venice is as simple as it gets: dump it in the canals and hope the tides carry it away.

It may sound overly simplistic, but it has worked for centuries (as long as the quantities of sanitary sewage was not too high). Brick tunnels called *gatoli* were built to bring the wastewater from buildings to the canals. These had to be periodically unclogged. The canals themselves needed to be periodically dredged to remove some of the sediment brought in by the tide. But, from the 1960s, the local government put too much faith in the *gatolis* and canals and decided to stop maintaining them, which led to a public health crisis. It did not help that, at the same time, water consumption (and therefore wastewater generation) increased significantly.

By the 1990s, the city's response was twofold. First, the *gatolis* and canals started to be maintained again, which is effective to get the wastewater dumped in the canals, but sanitary wastewater is highly polluted. The second solution is one that anyone who lives in an unincorporated or rural area knows: septic tanks. Septic tanks were built in many basements, especially in hotels and restaurants that generate a lot of sanitary wastewater. More than seven thousand tanks[23] were

---

23. I collected this value from a travel website. While I cannot judge the accuracy of the number, the website does nicely explain how sanitary wastewater is handled in Venice. See L. Romeo,

built (which sounds like a lot, but it does not include all buildings). After all, septic tanks are not magical. Generally, they work by applying the pre- and primary treatment processes covered in this chapter (i.e., greases float up, sludge sinks to the bottom), but the wastewater that is disposed of is still polluted. Finally, septic tanks do fill up and need to be periodically emptied. That is where the *pozzo nero* (cesspool) boats come in—even poop gets to ride on gondolas in Venice.

Overall, Venice is not a model example when it comes to sanitary sewer management. A new solution will be needed so that less polluted wastewater gets disposed of in the lagoon. In the meantime, if you ever get to visit Venice when it is flooded, I would not take a swim in the waters. No matter how fun it seems, that water should be contaminated, and you may bring back home a gift you did not wish for.

One last thing before we close this story: Learning about the sewer system made me wonder where Venice gets its fresh water for drinking. Surely, it must be imported from the mainland now (and it is), but where did it come from historically? Do you remember when I mentioned that every settlement had common features, including a church and a public square? Public squares (*campi*) were more than public areas. They were large reservoirs and treatment facilities for rainwater. First, they are all designed with a slight slant to bring all the rainwater to one or multiple inlets around the square. The bottom of each reservoir was filled with clay as an impermeable agent. The reservoirs were filled with stones and sand that filtered and brought the semi-treated water to open wells, where people could collect it. Again, Venice is extraordinary.

"Sewage in Venice: How Does It Work?" Best Venice Guides, February 28, 2020, available at https://bestveniceguides.it/en/2020/02/28/sewage-in-venice-how-does-it-work/? (accessed May 29, 2024).

CHAPTER FIVE

Stormwater Management

Rain or Shine, Climate Resilience in Copenhagen

THE LAST CHAPTER WAS DEDICATED TO SANITARY WASTEWATER THAT comes from humans and leads to the construction of sanitary sewers. The second type of wastewater that deserves as much attention is stormwater. The management of stormwater may not seem like an important personal issue—after all, we do not consume stormwater like we consume water or electricity—yet it has become a leading driver of the transformation of cities worldwide in the first half of the twenty-first century. In this chapter, we will learn about stormwater management by traveling to the Danish capital: Copenhagen.

To many, Copenhagen is famous for its Scandinavian lifestyle and crazy architecture (including a famous combined heat and power waste-to-energy plant that turns into a ski slope in the winter[1]). It is also famous for its statue of *The Little Mermaid* that forever sits on a rock with a face full of sorrow despite the awe it brings to every visitor. But do not go and associate Denmark with the words *small* and *quaint*. For a country of fewer than six million people, Denmark is home to a surprisingly high number of major international companies, including LEGO (yes, the plastic brick toy), Maersk (the shipping giant), Novo Nordisk (one of the largest producers of insulin in the world), Carlsberg (the brewing

---

1. Called Amager Bakke. If you have not heard about it, look it up.

company), and, oddly enough, Europe's largest pork producer: the Danish Crown Group. Copenhagen—and Denmark more generally—is full of surprises. Personally, I remember enjoyable walks through the Christiania neighborhood (look it up; it is an unexpected place in the middle of an international capital). And I hope that, by the end of this chapter, you will agree that Copenhagen should also be famous for its stormwater management. Before we get to know why, we need to know what stormwater management is.

First, we must look into the sky. Every single drop of rain is subject to three possible phenomena once it falls on the ground. First, it can fall on trees, buildings, or any other surface and stay there, either to evaporate or to be absorbed (e.g., by a plant): this is called *abstraction*. Second, it can fall and infiltrate the ground or pond on a surface until it eventually becomes groundwater: this is called *retention*. Third, if the drop of water cannot stick to a surface (no abstraction), or if the ground is full and ponding is not possible (no retention), it will simply follow the natural slope of the surface and flow: this is called *runoff*. In the field of hydrology, rainfall is technically referred to as precipitation, and, as just described, rain that comes from precipitation can turn into abstraction, retention, or runoff.[2] For our purposes, we are interested in the last one, runoff, as it can represent the worst nightmare for some city managers.

In nature, the roles of abstraction and retention are large. After all, plants need water, and water can easily infiltrate open soil. However, buildings, roads, bridges, and other infrastructure that make the physical city tend to be impermeable, and impermeable surfaces mean little to no infiltration. In cities, most of the precipitation turns into runoff. For small rain events, managing stormwater is fine. Most of the rain is abstracted and rapidly evaporates after the event. Through gutters, any runoff is channeled to the grates that we see at curbsides to disappear into the sewer; these grates are called *catchment basins*. For large rain events,

---

2. The equation is $P = I_a + F + Q$, where P is precipitation, I_a is initial abstraction, F is retention (i.e., infiltration and ponding), and Q is runoff. In physics, the symbol Q is often used to represent a flow—from flow of particles to flow of heat (in this case, we are looking for the flow of runoff).

however, the amount of runoff generated rapidly becomes overwhelming—because the runoff has nowhere to go and sewers are full—which can lead to flooding.

You might think that this is not a new problem. Cities have always received rain. In fact, early sewer systems were built purely to handle stormwater since people did not consume much water and used cesspools at home to handle sanitary wastewater, as we learned in Paris. Cities are therefore equipped with infrastructure to handle stormwater. But stormwater management has become an important problem in the early twenty-first century, primarily for two reasons. First, climate change is radically altering precipitation patterns, and, as a result, intense rainfalls tend to be both more frequent and more severe. Second, cities have grown a lot since the nineteenth century, which means that the size of impermeable surfaces has grown a lot as well, resulting in much larger flows of runoff than what stormwater sewers were designed to handle. This legacy (often undersized) stormwater infrastructure is therefore unable to do its job properly. The problem is even worse in cities that have combined sewer systems (sewer systems that carry both sanitary and stormwater sewers) since heavy rain events lead to combined sewer overflows (CSOs), during which raw, polluted wastewater flows to natural water bodies like rivers and lakes. CSOs have become much too common in cities that have combined sewer systems. The bottom line is that most cities worldwide are overwhelmed by the amount of stormwater they need to manage, and new solutions are needed.

Before we discuss new solutions, let us discuss how cities have traditionally managed stormwater. Historically, most cities were small and did not even have significant stormwater infrastructure. Stormwater was mostly managed by building ditches to drain any excess water, like runoff, to nearby surface water bodies, including rivers and lakes. In cities that faced recurring flooding problems, barriers and dams were often built to block overflowing rivers. Naturally, some exceptions exist, and some cities innovated more than others. Perhaps unsurprisingly, Rome was one early innovator with the construction of Cloaca Maxima, which is a large underground tunnel that channels wastewater to the Tiber River. The name Cloaca Maxima literally translates to "Greatest Sewer." It is majestic and still in use today.[3] Other cities, such as Tokyo in the seventeenth century, innovated in different ways.

Essentially, the strategy has historically been to get rid of the stormwater as quickly and efficiently as possible. This is also why, by the late nineteenth century, concrete or brick channels, which we now call *sewers*, were built underneath streets in many cities. Part of the reason to bury sewers is that the same fluid mechanic principles at work for water collection apply. Pumps are typically not used in sewer systems.[4] Instead, the channels are cleverly built in such a way as to create a downward slope in the direction where we want the wastewater to go. As a result, a sewer system consists of a giant network of carefully sloped channels so that all wastewater flows in predefined directions. Unlike water distribution, sewer systems tend to be tree networks—that is, the network does not form a grid but a tree-like structure. Again, that is because the conduits are always sloped to a lower elevation.[5] At the end of combined sewer systems you will generally find a wastewater treatment plant; separated sewer systems for stormwater will end at a surface water body.

The amount of wastewater carried out by sewer systems depends on the number of connections to sewer systems. As cities grow and new connections are added to the sewer system, the existing underground concrete and brick channels become too small to carry all the wastewater, especially during heavy rains. As a result, the system can fill up rapidly and overflow, thereby leading to flooding and, in combined sewer systems, to basement sewer backups (i.e., raw sewer backs up in basements). One way to remedy this dilemma is to put in new, larger underground channels

---

3. After all, the Romans did use a lot of water thanks to their extensive water collection system, and they had to find a way to deal with the resulting sanitary wastewater.

4. The exception is where cities are below the level of nearby surface water bodies, like New Orleans, or where parts of cities are at a lower elevation, like Paris. As a result, pumps must be installed at strategic locations to elevate the wastewater to a higher ground and where gravity can then transport it. Unlike water distribution, pressure should not build up in sewer systems.

5. That is not possible with grid networks. For water distribution, small changes in elevation do not matter since all pipes are pressurized.

that can carry larger flows. The problem with this solution is that it is extremely costly, since it requires digging up all city streets again, and sewer systems tend to be the deepest buried utility (below electric and gas lines, water conduits, and telecommunication cables). Some cities have decided instead to build massive underground tunnels to collect and store overflowing wastewater in various parts of the city until it stops raining, at which point the wastewater is pumped back up to be treated in wastewater treatment plants—this is what Paris and Chicago did.[6] All these solutions—typically called *gray* infrastructure because of the color of concrete—are extremely costly and tend to enjoy only limited success. Flooding and sewer backups still occur, albeit less frequently. After using the same strategy for years, if not millennia, we must revisit how stormwater is managed.

Becoming more popular in the early 2000s, the concepts of *green infrastructure* and *low-impact development* have emerged as potential saviors. If we go back to the fundamentals, we need to realize that runoff is generated because surfaces are impermeable. In nature, a high proportion of the rain is abstracted and retained instead because surfaces are permeable. In cities, while there is not much we can do to favor abstraction, by making surfaces more permeable, we can increase the potential for retention. Put simply, instead of getting rid of stormwater as quickly as possible, we can keep it where it falls and let as little of it as possible enter the sewer system. Favoring the *retention* and *detention* of stormwater—more on those terms later—requires a complete shift in how we design streets and stormwater infrastructure. Another term used for green infrastructure and low-impact development is nature-based solutions, which try to replicate proven methods that are present in nature. Now, let's return to Copenhagen.

From central Copenhagen, you can ride public transport or walk about one hour north to the neighborhood of Østerbro. Thanks to a series of stormwater management projects, Østerbro has become known as a

---

6. The tunnels in Paris can store about one million cubic meters of wastewater.

climate-resilient neighborhood (or *klimakvarter* in its original Danish). The end goal in Østerbro is to adapt and be ready to face the impacts of climate change, with a particular focus on being able to handle heavy rain events—no flooding, no backups regardless of how intense a rain event may be, just business as usual, and all of it through seamless, silent, carefully engineered infrastructure. To reach this goal, the Østerbro neighborhood divided land use into four types: urban spaces (large common areas), courtyards (greenspace in private properties), streets, and buildings. An important goal of each of these types is to hold stormwater in place for as long as possible as opposed to sending it to the sewers.

To be more climate resilient, private properties can be retrofitted to require less energy and less water, but also to generate less runoff. Buildings can be equipped with green roofs (e.g., gardens on roofs) that soak in and retain water during rain events. In courtyards, rain barrels, which fill up when it rains, can be installed at the outlet of downspouts. Another strategy is to build *rain gardens* at the outlet of downspouts. Dig a hole, fill it with rocks, add some soil and plants with long roots, and you have yourself a rain garden. The rocks essentially create some void space that store stormwater when it rains, and the long-rooted plants help hold the water in place. Rain gardens can be small, taking up little space in a front or backyard—anybody can build one—or they can be large and spread across a courtyard. Well-engineered rain gardens directly contrast with grassland. Grass is compact and has small roots that offer no space for infiltration and cannot hold much water. After a bit of rain, grass quickly gets soaked and muddy. Additional rain immediately turns into runoff. Grass is known to perform poorly in stormwater management. Instead, selecting long-rooted plants that can hold water is much more effective.[7]

Away from downspouts, footpaths in courtyards can be equipped with permeable pavers so that rainwater can trickle down the void space between them, and different sizes of gravel can be used underneath the

---

7. In North America, we talk about *native plants* since grass is an import from Europe. Native plants have long roots to store and retain rainwater when it rains a lot in the spring so they have ample water during typically dryer summer months.

pavers to offer some void space to store stormwater. In Østerbro, the Klimakarré building complex features several of these strategies.

Despite these efforts, during heavy rains, green roofs, rain barrels, rain gardens, permeable pavers, and any other LID strategy can fill up, overflow, and generate runoff. It is time to turn to the streets.

In Østerbro, many streets such as Tåsingegade are turned into "cloudburst" roads. A *cloudburst road* is defined as "a public road that can carry cars, cyclists and pedestrians, but at the same time is able to function as a channel conveying the rainwater in torrential downpours safely away from the neighbourhood."[8] The main principles applied in buildings and courtyards are similar to those of streets: keep the water in place (e.g., by building *bioswales*) and channel any runoff away in a specific direction. Bioswales are channels that are filled with gravel or rocks and long-rooted plants. They are a mix of rain gardens and traditional drains. Like rain gardens, they can store stormwater, but runoff is still generated during heavy rains, and, as a result, bioswales are sloped to channel excess stormwater where we want the stormwater to go. And where do we want excess stormwater to go? To urban places, of course.

Urban places are large common areas like roundabouts, plazas, or even parks. Here, we now have even more space to store stormwater, often in *retention* and *detention basins*. Essentially, both are storage facilities, but retention basins have open bottoms and can let water infiltrate the ground, whereas detention basins have closed bottoms. For example, a pond that receives stormwater is a retention basin. In contrast, an empty tank that is built underground to store stormwater before it is sent to the sewers is a detention basin. Rain gardens and bioswales are examples of retention basins (the term *bioretention* is even used). Catchment basins built at street curbs are detention basins. It is easy to tell when a catchment basin is full since water starts ponding at street curbs.[9] You may wonder why

---

8. For this quote and for general information on the Klimakvarter neighborhood, see https://klimakvarter.dk/ (accessed May 29, 2024).

9. Oftentimes, street curbs are flooded on purpose. The ponding of water at street curbs during heavy rains that makes you jump over them does not mean that there is a backup but that the grate is working as intended. Essentially, the goal is to slow down the rate at which stormwater enters the

detention basins are used at all since infiltration always seems more beneficial, but that is not the case. Even in the ground, water always follows the path of least resistance, which sometimes leads to the foundation of a nearby building that may get flooded (if there are cracks) or that may affect the structural property of the soil providing support for the building. Again, the adjective that gets repeated is "careful."

Østerbro features at least two urban spaces: Tåsinge Plads and Skt. Kjelds Plads, which are about two hundred meters apart. Tåsinge Plads used to be a small park covered with grass, and Skt. Kjelds Plads used to be a typical roundabout with small trees. Although both urban spaces used to feature greenery, the landscaping did not favor stormwater retention—that is, there was no space for the stormwater, and, as a result, the stormwater simply ended in the sewer. Look around in your city; I bet you will easily find sidewalks with greenery. Because the sidewalks are elevated, the stormwater flows toward the gutter and is channeled to the curb to enter the sewer instead of being directed to the greenery to be used by the plants.[10] In Østerbro, the small park at Tåsinge Plads was extended to incorporate unused asphalt, where an all-grass area was turned into a "luxuriant 'Danish rainforest.'"[11] Footpaths and a seating area were built and covered with permeable pavers. A similar vision was adopted for Skt. Kjelds Plads.

In the end, urban spaces often offer the perfect venue for stormwater management. In fact, many cities use urban spaces as small artificial floodplains. As a result, entire parks or outdoor sports facilities are flooded on purpose during heavy rains, which is preferable to flooding streets and buildings. This is why many neighborhoods that are destroyed during severe storm events (e.g., after superstorm Sandy in the New York area in the United States) get turned into parks that can be flooded on purpose during the next severe storm.

---

sewer system. In that case, the street curb itself becomes a detention basin, and it is preferable to have streets partially flooded than sewers backing up in basements.

10. And the plants must be watered as a result, even though water often literally flows next to or under it.

11. Term used on the project website: https://klimakvarter.dk/ (accessed May 29, 2024).

Remember, the goal now is to be able to store stormwater through careful landscaping that allows enough storage space and controls where overflowing stormwater goes. Again, the adjective "careful" is important here. It is not just about planting a few plants. It is about designing green infrastructure (GI) and low-impact development (LID) to be able to store predetermined volumes of water that are released at an anticipated rate. Two variables are particularly important: flow rate and volume. Flow rates during intense rains may be so high that stormwater does not have time to infiltrate the ground, even if there is free space. GI and LID are designed to accommodate specific infiltration rates. Moreover, large rains generate a lot of stormwater, and volume must be considered, too. Generally, we speak in terms of *return periods* and *duration*. For example, a design should be able to accommodate a twenty-four-hour, one-hundred-year rain—that is, it should be able to accommodate a rain that is most likely to happen once in one hundred years with a certain intensity for a duration of twenty-four hours. The term *one-hundred-year rain* is confusing, as people assume such a rain event can only occur once every hundred years, but that's not it. It means that it has a 1 percent chance of happening in any year, and it is not uncommon to have several one-hundred-year rains over a period of a few years, especially as an effect of climate change. Most cities have the necessary data; as a result, GI and LID can be engineered to respect certain design specifications such as a predetermined return period and duration. Even if these values evolve with climate change,[12] we have a general idea of the performance we want out of our design. And yet we must still acknowledge that even larger storm events might happen that can flood a city regardless of how well a GI and LID performs, simply because more rainwater pours down than what was designed for in

---

12. With climate change, what used to be a one-hundred-year storm event, for example, might become a fifty-year event (i.e., having a 2 percent change of occurring). These statistics tend to be made by collecting data over long periods of time, but because climate change is continuously altering rain patterns, these values also continuously change. Recalibrating these values is not sufficient since they would have to be continuously recalibrated. What can be done is to increase regulations. For example, instead of accommodating a one-hundred-year rain, we might ask to accommodate a five-hundred-year rain (i.e., 0.2 percent chance of happening). This solution is not perfect, but we need a means to estimate a volume and flow rate, and this method seems as effective as any other.

the first place. This is why every city must think seriously about stormwater management and elaborate an effective plan that uses a combination of green and gray infrastructure.

We have also omitted an important argument here in favor of GI and LID. Beyond stormwater management, GI and LID are known to generate myriad benefits. They help promote local biodiversity, sequester carbon, and treat stormwater, to name a few. They are also known to generate economic and social benefits to a community. All these benefits are often termed as *ecosystem services* (which is a little bit of an obscure term to my taste to describe many wonderful benefits). Overall, this is why I mentioned that GI and LID are leading drivers of the transformation of cities in the first half of the twenty-first century. Cities worldwide are adopting GI and LID strategies, and the few strategies that we have discussed here represent only the tip of the iceberg. GI and LID strategies can be implemented in every single street, courtyard, and park, greening up the space and generating myriad benefits beyond stormwater management. Who could have guessed that a usually undervalued area such as stormwater management would become such an important driver of change in the world?

After our in-depth exploration of water, we now move on to another facet of the world of infrastructure. It is time to learn about transport, starting with traffic and roads. What better place to learn about traffic and roads than a city sometimes referred to as the City of Highways: Los Angeles?

BEING SMART IN KUALA LUMPUR

Kuala Lumpur—or simply K.L., as many call it—is the capital of Malaysia. K.L. has become a major financial center with a dense financial district that proudly features the Petronas Tower(s),[13] which bolster beautiful Islamic designs that

---

13. I put the plural *s* in parentheses because I remember thinking that the whole complex made a single entity when I visited K.L. Plus, the two towers are linked by a sky bridge—does it mean there is one tower or two? I like to think one only.

can be appreciated from the large and lush KLCC Park, situated in the middle of a modern neighborhood. The city is composed of a mix of old and new buildings, manifesting the rich history of the region. K.L. also has an amazing food scene.

Geographically, K.L. is three degrees north of the equator. It has a tropical rainforest climate, which means it is hot and humid year-round, and it rains a lot, especially during the rainy season. One thing we learned in this chapter is that stormwater requires space. What the GI and LID strategies discussed here provide is void space for stormwater that can be distributed over a neighborhood or even an entire city. In a tropical rainforest climate, it means that lots of void space is needed. Rain gardens and bioswales can help, but they are often insufficient. In a tropical rainforest the soil is porous, and enormous, lush, green trees and other plants absorb that stormwater to grow. In a city paved with concrete and asphalt, stormwater quickly turns into runoff.

K.L. faces another big problem that has plagued most modern cities: it suffers from heavy traffic congestion. To most engineers, stormwater and heavy traffic are completely separate problems that require completely separate solutions, but the people in K.L. see it otherwise. Could there be infrastructure that can help address both problems simultaneously?

When surfaces are impermeable, dealing with large volumes of stormwater requires large detention or retention basins. But this infrastructure is needed only when it rains. Could it be used to alleviate traffic when it does not rain? This is exactly the solution that was adopted. The Stormwater Management and Road Tunnel (SMART) project is a ten-kilometer-long underground tollway that crosses part of the city and is used to carry traffic. But when it rains, the tunnel is closed. It becomes a large detention basin that is flooded with stormwater.

The tunnel has three levels. The bottom level carries only water, and it is often sufficient to handle small rain events. The two top levels carry vehicles at most times, but they are closed during heavy rains. The project comes from a collaboration between the Malaysian Highway Authority and the Department of Irrigation and Drainage Malaysia—an unlikely collaboration in many countries. Construction for the project began in 2003, and the tunnel opened in 2007.

I would not call the SMART project an LID, but it offers some of the void space necessary to deal with torrential rains. And when it does not rain and is not flooded, it can be used to carry traffic. Usually, I am not a fan of tunnels for stormwater management. Also, as we will soon discover, building more road infrastructure in already overdeveloped cities that suffer from heavy traffic congestion is seldom the solution. But there is something about the SMART project that makes sense as a semi-temporary solution. I write *semi-temporary* because I do not think K.L. should build more SMART projects. Yet I commend K.L. for building multifunctional infrastructure, which is, to me, one of the most important shifts that needs to happen in the engineering culture (engineers tend to be too siloed,[14] limiting their creativity). Departments and competencies have become so isolated from one another that new infrastructure systems are often built with one single function in mind. But it does not need to be like that.

Driving through a tunnel is rarely something that attracts tourists, but I remember my SMART experience clearly. It was close to 7:00 a.m. My wife and I had to go to K.L. International Airport to travel back to Vietnam in fall 2019. I still had not seen the SMART project and felt a bit frustrated about it. This was my chance. We hailed a taxi and asked to go to the airport, but I had one condition: we had to drive through part of the SMART tunnel. The driver accepted. We left the hotel. After some time, I could see some street signs pointing to the SMART tunnel. The car then engaged on a downward ramp, and there we were, in the tunnel. I was so happy. The tint of the tunnel was somewhere between yellow, orange, and brown. The tunnel was smaller than I expected at first, but then I realized we were driving on the upper floor since I could see the rounded ceiling. The pavement was damp, which made me think that the tunnel might have been flooded a few hours before—in fact, we had a severe thunderstorm the previous night. There was no traffic. It was only us and the infrastructure. After about ten minutes, we engaged on an upward ramp and continued to the airport.

If you ever get to visit K.L., make sure to have tasty beef noodle soup and take a ride through the SMART tunnel, especially a day after it has rained when the tunnel walls and roads are still wet.

---

14. If you happen to meet an engineering student, do not ask them which area they plan to specialize in (e.g., environmental versus transport versus structural in civil engineering). Instead, ask them which specialties they plan to integrate to develop creative solutions. (If you are an engineering student already, apply this question to yourself.)

PART TWO
TRANSPORT

CHAPTER SIX

Traffic and Roads

Stuck on the 405 in Los Angeles

"STUCK ON THE 405." I HAVE HEARD THIS LINE COUNTLESS TIMES IN movies and TV shows. Interstate 405 is one of the many expressways that spread across the Los Angeles region like a spiderweb. Starting north, in San Fernando, the 405 crosses the Santa Monica Mountains, going by the Getty Center, and continuing through West Los Angeles. It then bypasses LAX (the primary Los Angeles airport) and continues south through Inglewood and Long Beach, going all the way south to Irvine. In total, the 405 spans 116 kilometers (72 miles). At free-flow speed, it takes about an hour to drive from one end to the other, zooming through some of the wealthiest and poorest areas of the Los Angeles region. The 405 is infamous for being the busiest and most congested expressway in the United States. I remember driving on the 405, being stuck in traffic, and feeling like it was part of the local experience. Despite its seven lanes in each direction—a total of fourteen lanes—I was stuck in traffic. Don't you think this is absurd?

Los Angeles (known by its two-letter abbreviation L.A.) makes people around the world dream. Hollywood has been the movie and television capital of the world since the 1930s. Every tourist visiting L.A. is constantly on alert, trying to spot a celebrity. L.A. is also famous for its perfect weather. It does not rain often, and the temperature always hovers around 25°C (77°F): just perfect. For urban planners and enthusiasts,

L.A. is also known as a giant suburb and a traffic nightmare—a textbook example of what not to do. L.A. might be the city of angels, but Angelinos[1] do not fly across their city. Instead, they are stuck in traffic. In 2022, they spent 122 hours on average stuck in traffic;[2] that is over five full days, the highest of all U.S. cities.

After a journey through the realm of urban water, it is time to discuss transport.[3] Like we did with water, we will look at various elements of transport over the next several chapters, but we will start with traffic and roads. Roads predate cars, but cars have had such a disruptive impact on how we build and run infrastructure in cities that they deserve to be discussed first. It is difficult to think of a better city to represent traffic and roads than L.A., a city that was built for cars.[4]

Unlike other infrastructure systems, transport is less seamless and silent. As I have heard someone say once, "Everyone is a transport expert," since we all experience it daily, and we often complain about it.[5] Our goal first will be to learn some basics of traffic flow theory.

When we travel, the personal objective of a user (i.e., you and me) is usually to *minimize travel time* so that it takes as little time as possible to travel from an origin to a destination. In traffic engineering, this *minimize travel time* objective is converted to something different: *maximize flow*. Essentially, the objective is to have as many vehicles as possible flow through a road—another word for *flow* is *throughput*. Therefore, regardless of individual travel times, the objective in traffic engineering is to

---

1. Angelinos are the people of Los Angeles, like New Yorkers for New York City.

2. D. Schrank, L. Albert, K. Jha, and B. Eisele, *2023 Urban Mobility Report*, published by the Texas A&M Transportation Institute with cooperation from INRIX (2024).

3. In the United States, the term *transportation* is generally used instead of *transport*. The two terms are synonymous. We will use *transport* simply because it is more commonly used internationally; a quick Google Trends analysis showed that *transport* is used about three times more than *transportation*. Moreover, the fact it is shorter also means it requires less energy to store digitally, and it saves a lot of paper. See a blog post from a famous transport engineer, David Levinson, about this subject: https://transportist.org/2011/04/01/ (accessed May 19, 2024).

4. The movie *Who Framed Roger Rabbit* offers an entertaining account of the development of expressways in Los Angeles.

5. And often thinking that *we* have the solution. I cannot count how many times I have heard someone say something along the lines of "Why don't they deal with traffic like *that*? It is common sense." Trust me, it is not common sense. Yet a solution does exist, and we will discover it over the next few chapters.

maximize the number of vehicles that pass through, not to minimize travel time. Although different, the two objectives are directly related.

Transport flow is not unlike the flow of water.[6] In fact, a little like in stormwater management—in which, historically, the strategy has been to get rid of stormwater as rapidly as possible—the strategy in transport has also been to "get rid" of the cars as rapidly as possible. In transport terms, the strategy has been to allow as many vehicles as possible to flow through a road.

So what is flow? Water flow is expressed as a volume of water per unit time (e.g., in liters of water per second). Transport flow is expressed in vehicles per unit time; our basic unit is therefore the *vehicle* (even if trucks and SUVs are larger than cars). A common unit of flow in traffic engineering is vehicles per hour. In our daily life, we are more accustomed to *speed* that is expressed in distance traveled per unit time (for example, in kilometers per hour). Flow and speed are in fact related. If flow is in vehicles per hour and speed is in kilometers per hour, there should also be another variable expressed in vehicles per kilometer. This other variable measures how packed a road is—few cars versus many cars—and such a variable, called *density*, does exist. In the end, the relationship is the following: flow (vehicles per hour) is equal to density (vehicles per kilometer) multiplied by speed (kilometers per hour).[7] Put differently, we take the number of vehicles we have on a road and multiple it by the average speed[8] to know how many vehicles we expect to flow through the road.

---

6. Most engineering fields tend to have similar variables like *flow*, *velocity*, or *power* because most engineers have an educational background in physics. It is therefore quite natural that early traffic engineers considered vehicles like particles. Now, does it make sense? I am not sure; we have essentially removed the human element so that real-world phenomena fit into simple equations. Concepts from physics have pervaded most scientific disciplines, from engineering to economics. I would love to see a different approach to traffic engineering—for example, why not an approach from biology (start from the complex and seek to understand the simple)?

7. In equation form, we have $q = k \cdot v$, where q is flow, k is density, and v is speed. Looking at the units, we can see that [vehicles/hour] = [vehicles/kilometer] × [kilometers/hour].

8. There are at least two ways to measure speed. The one we are accustomed to is time-mean speed, in which we measure the average speed of all the cars passing a point. In transport, we prefer to use the time a vehicle takes on average to travel a certain distance (e.g., the time to reach downtown from the airport). This is called space-mean speed. When we go through the math, we find that space-mean

Now that we have defined what flow is, we need to figure out how to maximize it. What the flow equation tells us is that to get a high flow, we must have a high density and a high speed. Ideally, we would have a road packed with cars that are driving extremely fast. As virtually everyone has experienced, this is not possible; during a traffic jam, speed is low.[9] In fact, there is a trade-off between speed and density. To reach free-flow speed (i.e., the maximum allowable speed on a road), density should be low, which means that flow is low as well. The opposite is when the jam density is reached (i.e., the maximum density happening during a traffic jam), at which time speed is nearly zero, also resulting in a low flow. There is a sweet spot, somewhere in the middle, where density is fairly high but not too high, which allows speed to be fairly high as well, although not as high as free-flow speed. This maximum flow is also called the *flow at capacity*—that is, when the full capacity of the road is utilized.

These observations are best captured by a famous graph in traffic engineering called the *fundamental flow diagram* or *fundamental diagram of traffic flow* that plots the relationship between flow, speed, and density. It has a total of three plots: speed versus density, flow versus density, and flow versus speed. The plot of speed versus density looks like a downward slope: when speed is high, density is low, and vice versa. The two remaining plots—flow versus density and flow versus speed—are parabolas. This shape expresses the fact that when speed or density are low or high, the flow must be low. The apex in the middle achieves the maximum throughput, which happens when both speed and density are neither too high nor too low.

Figure 6.1 shows two perfect parabolas. It is called the Greenshields model.[10] The real shape of the curve when we collect traffic data is

---

speed is the harmonic mean of speed (as opposed to the arithmetic mean). You can learn more about it in my textbook *Urban Engineering for Sustainability* (Cambridge, MA: MIT Press, 2019).

9. We will not discuss autonomous vehicles (AVs)—driverless cars—here, for which achieving high density and high speed is maybe technically possible since we can remove human error from traffic. In a sense, AVs may be the particles that transport engineers have modeled all along.

10. After the American transport engineer Bruce Douglas Greenshields, who developed and published his famous model in 1935.

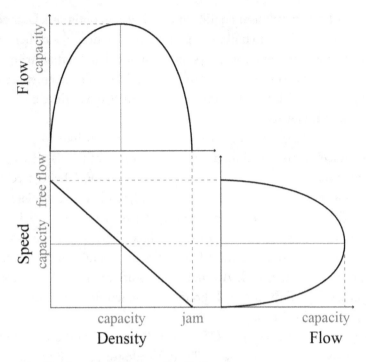

Figure 6.1. *Fundamental diagram of traffic flow.*

different. The righthand side of the parabola extends much farther, and the apex occurs somewhere between 20 and 25 percent of the jam density.

We can now understand why minimizing travel time and maximizing flow are different. Namely, more vehicles can go through when density is not low (i.e., when there are other vehicles around), and a not-low density translates into a speed that is lower than the free-flow speed. Although we prefer to drive at free-flow speed to minimize our personal travel time, we are not alone on the road. Other drivers want to travel as fast as possible too, but it is only when all of us drive a little slower that we can collectively have shorter travel times. This is why traffic engineers focus on flow. In practice, to model traffic, traffic simulators—computer programs used to simulate traffic on a road system—obey two principles called

Wardrop's principles:[11] user equilibrium and system optimum. User equilibrium dictates that, at equilibrium, no user can reduce his or her travel time by taking a different route. System optimum dictates that, at equilibrium, the total travel time of all users is minimized. We therefore see that although they are different, minimizing personal travel time and maximizing flow are related.

To make things a little easier, traffic engineers like to grade the performance of individual roads at different times of the day. This grade is called level of service (LOS). It ranges from A to F, A being free-flow conditions with virtually no vehicles around, while F is complete congestion (i.e., bumper-to-bumper traffic, the road is over capacity). At night, most roads have an LOS of A since there are not many cars around. During rush hour, the LOS of a road depends on how much it is used. For example, in L.A., many segments of the 405 expressway have an LOS of F during rush hour. Maximized flow generally means an LOS around C or D.

Roads with a consistent LOS of A or B are underutilized, and roads with a daily LOS of F are overutilized. While we generally do not care too much about roads that are underutilized, ideally, we want to ease up the congestion of roads that are overutilized. In fact, not only is overutilization bad in terms of user experience (nobody likes to waste time being stuck in bumper-to-bumper traffic), but it also causes the pavement to wear more rapidly. As a result, more frequent maintenance is required, which is costly and affects traffic, further impacting users.

So, what is the solution when the LOS of a road is rated F? There are not one or two but at least three solutions.

We have not discussed road capacity and number of lanes at this point. From small alleys to large expressways, not all roads have the same capacity. Based on size (i.e., number of lanes) and speed limits, some roads have a small capacity, and others have a large capacity. The first solution is

11. After the British mathematician John Glen Wardrop, who developed the two principles in 1952.

to increase capacity by widening existing roads or by building new roads. Adding new lanes to a road increases the jam density. Adding new roads increases the capacity of the entire road system. Widening and adding new roads was the go-to strategy for most cities in the United States after the 1950s, and it resulted in these familiar webs of urban expressways that include the 405.

The second solution does not apply to the 405 because it is an expressway, but it has been implemented in virtually every city in the world: traffic signals. With their universal three lights—red, yellow, and green—the goal of traffic signals is to control traffic so that it is both more fluid[12] and safer. In the United States, the *Manual on Uniform Traffic Control Devices* (MUTCD) regulates the use of traffic signals; the official terms given to green, yellow, and red lights are *right-of-way*, *yellow change*, and *red clearance intervals*. Traffic signals can be either *pretimed* or *actuated*. Pretimed means that signal timings are fixed, regardless of demand, although the actual timings may change during the day to accommodate an expected traffic (e.g., rush hour). Actuated means that signal timings vary depending on demand. The typical case is when a pedestrian presses a button to cross a street. On side streets that carry little traffic, loop detectors are also sometimes installed in the pavement so that the green light is turned on only when vehicles are detected. So-called intelligent transport system (ITS) solutions have also been used extensively to adapt signal timings to real-time demand. Optimizing each individual traffic signal in an entire system in real time is practically impossible;[13] instead, usually, a pool of scenarios is preprogrammed and activated to accommodate a real-time demand. The most famous ITS solution to control traffic signals is probably the Sydney Coordinated Adaptive Traffic System. Originally

---

12. *More fluid traffic*: here is another illustration that comes directly from physics, comparing traffic to a liquid.

13. Running complex optimization algorithms on large road systems requires a lot of time and computing power. Running these algorithms constantly and adapting traffic timings in short time intervals is practically impossible (at least as of this writing). And even if it were, the time gains might not be significant. It is preferable to turn to the third solution.

developed for Sydney, Australia, the solution has been implemented in many cities around the world. In the end, if done properly, for given traffic conditions, traffic signals can help better utilize roads. As a result, traffic can be more fluid altogether without widening existing roads or building new roads.

The fact that the 405 has an LOS of F during rush hour should raise a red flag. In L.A., the first solution to increase capacity by adding new lanes and new roads has been implemented to its full extent over the years. Moreover, L.A. also developed its own ITS traffic solution to adjust traffic signal timings in real time—the Los Angeles Automated Traffic Surveillance and Control (ATSAC) system—and it works well. A third solution examines the problem from a completely different angle.

The problem is that, for the longest time, L.A. (but really most cities) considered only one part of the equation: *supply*. By considering capacity only, the authorities overemphasized the *supply* and *utilization* of roads, without thinking about the *demand*. Demand is how much the road is utilized by individual users. It is expressed by two main characteristics: the number of users and the length of trips (the distance between an origin and a destination). We know that demand varies during a typical day (rush hour versus night traffic), but it also varies over the years. As the regional population increases, so does the demand. Moreover, with more roads, people and businesses can relocate farther away, generally resulting in higher trip lengths. No matter how many roads are built or how many new lanes are added, demand catches up with the supply, resulting in traffic jams. In economics, this phenomenon is known as *induced demand* (or *induced traffic* in our case). The great urban historian Lewis Mumford famously said in 1955, "Building more roads to prevent congestion is like a fat man loosening his belt to prevent obesity." To get back to what I wrote earlier, don't you think that the presence of systematic traffic jams on a fourteen-lane expressway in a city full of expressways is absurd? The first two solutions to increase road capacity and utilization are not complete solutions, or at least they are effective only when demand is managed. We need to turn to a third solution that instead focuses on demand.

If a road system is overutilized, instead of increasing capacity, a more effective solution might be to reduce the demand for roads in the first place. We alluded to this solution in the water collection chapter in Rome. As mentioned above, demand is characterized by two things: number of users and trip lengths. By reducing either of the two, demand is lowered. To lower the first, some users need to either take fewer trips or change their travel mode by walking, cycling, or riding transit instead of driving. To lower the second, destinations need to be closer to their origins. A suite of strategies has been developed to lower demand. They are typically grouped within an all-encompassing term: travel demand management (TDM). TDM strategies to lower the number of users tend to be applicable in the short to medium term; they include increasing parking prices, offering effective transit alternatives, promoting car sharing, and adding a toll to an existing road. TDM strategies to shorten trip lengths tend to be applicable in the medium to long term, and they especially relate to land use. In particular, cities can adopt mixed land-use policies so that every neighborhood has residences, businesses, schools, shops, and so on, enabling people to carry out short trips to get what they want instead of having to drive long distances.

TDM strategies tend to be much more effective than increasing road capacity, but they were ignored for the longest time. Similar to situations of stormwater management, traffic engineers needed to completely change their objective or forget about trying to maximize flow.

Even L.A. has embraced TDM. In particular, downtown L.A. has transformed itself, adopting denser, mixed-land use practices. After years of constant suburban extension, L.A. has followed suit and built a dynamic downtown core. The city has massively invested in public transport as well since it offers the main motorized alternative to private vehicles. Being able to plan and run an effective public transport system is challenging, however. To learn more about public transport, let us cross the Atlantic Ocean and visit the city that built the first ever metro system: London.

A TALE OF TWO ELEVATED EXPRESSWAYS

Out of all types of expressways, I would argue that the worst is the elevated one. Like all expressways, elevated expressways lead to massive traffic congestion and poor air quality. Like all expressways, they tend to divide entire neighborhoods or cities into two, becoming physical barriers. But unlike other expressways, they block the view between two neighborhoods, often deterring movement between them. Anyone who has ever crossed an elevated expressway underpass, especially by walking, will immediately understand what I mean.

In many parts of the world, the 1950s, 1960s, and 1970s offered unprecedented economic growth. This time coincided with the advent of the new transport revolution—the private car—that had reached some maturity and was ready to be deployed massively. Cities were scrambling to make way for cars. Neighborhoods were destroyed. People (generally low-income and marginalized) were displaced. Giant structures of concrete and asphalt were built. Expressways were seen as engineering marvels. Along with skyscrapers, they were thought to represent the best of what modernization had to offer. But modernity in the age of the machine has a way of biting back. Everywhere they were built, expressways virtually destroyed the neighborhoods along them. Air quality worsened. Green spaces were replaced with gray, concrete ones. And the promises of express commutes never materialized because expressways rapidly filled up with cars. The star of this chapter, L.A., offers the best example.

In response, throughout the world, many cities have decided to remove their expressways, especially elevated ones. I would be surprised if elevated expressways still existed by the end of the twenty-first century. What follows is the story of not one but two cities that struggled with their elevated expressways. Both wanted to get rid of them, but one city did much better than the other.

On the other side of the country, away from California's sunshine and movie stars, lies the northeastern city of Boston, Massachusetts. The defunct Atlantic Avenue Elevated rail line was originally built in the 1920s and offered the perfect space to build an elevated expressway. The rail line was demolished in 1942, and the scrap metal was used for World War II. In 1951, construction of the first elevated expressway in the United States started. It was called the Central Artery (Interstate 93). When it opened in summer 1954, it was seen as an engineering marvel and a model of urban planning that would "smash the city's traffic bottleneck." It was even equipped with the latest technologies—on/off ramps that could melt snow. The mayor of Boston at the time, John Hynes, announced, "For twenty years we have dreamed and hoped for something like this." By 1954, only the north section of the expressway was open, but thanks to its six lanes, a newspaper article predicted that a twenty-five-minute commute would be reduced to only two minutes. By 1959, the Central Artery was fully built. It was 2.4 kilometers (1.5 miles) long, starting in Charlestown to the north, crossing the Charles River and landing in the North End before traversing Government Center, the Financial District, and Chinatown, all the way to the South End.

The expressway quickly turned into a nightmare. Congestion was so bad that the expressway earned several nicknames, including the "distressway" and the "largest parking lot in the world." By the 1990s, the Central Artery was congested for more than ten hours per day. Soon after it opened, plans were drawn to replace it, but it took forty years before the elevated expressway was finally demolished. But what do you replace it with? With the United States having such a car-centric culture, it was inconceivable to remove it without adding road capacity somewhere. Wouldn't congestion become even worse otherwise? Unsurprisingly, the solution that was adopted was to build a longer, larger expressway, but underground. The whole project soon gained a fitting nickname: the Big Dig. After many cost overruns, the Big Dig finally opened in 2008. At the street level, where the elevated expressway once stood, a linear park was built (still with three lanes on each side). Look it up on a satellite map, I think it looks like a scar as a reminder of Boston's dirty past. The important question may be: Did the project deliver? Is Boston free of congestion? A report from INRIX found that in 2023, Interstate 93 (the Big Dig) was the seventh most congested road in the United States.[14] The result is not surprising. Beyond induced demand, it is physically impossible to pour a large number of cars through a small downtown, no matter how large the roads are (think of a funnel).

Let's see how the other city approached its elevated expressway problem, this time from across the globe. After the war that divided the Korean peninsula, Seoul grew at an amazing rate. The population increased from 1.3 million people in 1953 to 6.2 million in 1973. Many families bought a car. Traffic congestion was bad. Space was needed to make way for the future. Here comes the elevated expressway. In Seoul, the Cheonggyecheon Stream offered the perfect place. The people there were poor and lived in insalubrious conditions. The stream was buried and covered in concrete, and by 1976 an impressive double-story expressway opened. New buildings were built around the highway. The project was seen as a great example of modernization. As you might expect, the traffic worsened quickly. The noise level was unbearable. The air quality was unhealthy. What was once a quiet stream turned into a traffic nightmare. In 2001, a Seoul mayoral candidate made it a central campaign promise to demolish the expressway. He easily won. By 2005, the expressway was gone. The stream was restored and turned into a linear park as well, similar to Boston. Unlike Boston, the park is below ground level, which means that it is particularly quiet and pleasant. But the real innovation is that the park is closed during rain events to be flooded with stormwater (to divert it from flooding the streets). I like to say that, in Seoul, a transport problem provided a stormwater management solution. The Cheonggyecheon River (as it is

14. B. Pishue, *2023 INRIX Global Traffic Scorecard with Q1 2024 Update*, INRIX Research (2024).

now called) has become my favorite infrastructure in the world, and walking along it is one of my favorite activities.

And what about congestion, you might ask? Has traveling around Seoul turned into utter chaos? No. People adapt. Demand changes. One study found that traffic initially got worse but soon after got better as people adjusted their departure times, switched routes, and changed travel modes.[15] The local government has also built one of the world's best transit systems. One report suggests that bus ridership increased by 15.1 percent and metro ridership increased by 3.3 percent between 2003 and 2008.[16] Property values increased significantly. Small-particle air pollution decreased by 35 percent, which means not only better air quality but also a lower ambient air temperature of about 5°C (10°F) on summer days compared to nearby areas. The local biodiversity also increased by 639 percent. These are amazing statistics that should inspire every city to come up with creative solutions.

Focusing on traffic, there is also an obvious reason why public transport is a superior travel mode in cities and why everyone should advocate for massive investment in public transport, even (or maybe especially) people who hate public transport. It is time to travel to London and learn more about it.

15. J.-H. Chung, K. Y. Hwang, and Y. K. Bae, "The Loss of Road Capacity and Self-Compliance: Lessons from the Cheonggyecheon Stream Restoration," *Transport Policy* 21 (2012): 165–78.

16. H. S. Kim, T. G. Koh, and K. W. Kwon, "The Cheonggyecheon (Stream) Restoration Project—Effects of the Restoration Work," Cheonggyecheon Management Team, Seoul Metropolitan Facilities Management Corporation. Seoul, South Korea (2009).

CHAPTER SEVEN

Public Transport

Minding the Gap in London

AT 6:00 A.M., ON JANUARY 10, 1863, A GASLIT STEAM-POWERED LOCOMOTIVE left Paddington Station in London and traveled six kilometers (3.75 miles) east, under the streets of London, to Farringdon Station, serving six stations along the way. Thus, the London Underground was born. It serviced 38,000 people on that day. Commonly known as "the Tube," it was the first underground urban rail system ever built.

We need to put this story in context. While many cities around the world now have underground rail systems, this was not the case in 1863. Back then, cars did not exist, the main modes of transport were walking and horse-drawn carriages, and electricity had not become a common utility yet—the London Underground was steam-powered in 1863. This must make you wonder: What could push people to want to build an expensive underground tunnel to run locomotives that carry people back and forth beneath streets that already exist and that already carry these people? A tunnel is built to overcome an obstacle such as a river or a mountain, but here, there were no such obstacles. In fact, infrastructure already existed—roads—that served the same purpose: to transport people. So, how does it make sense? The answer is all too simple. The rationale for building an underground tunnel beneath streets was solving a problem that has always plagued large cities and that has provided the resources for the most extravagant transport projects ever built. The answer is none other than our old friend: congestion.

In this chapter, we will focus on public transport, also called *public transit* or just *transit* for short. By definition, public transport is a means of travel that is not owned by riders (unlike private vehicles), regardless of whether the system itself is publicly or privately owned, and that is generally used by a large number of people; this is why public transport is also called *mass transit*. We will focus only on urban public transport systems, but airlines and intercity rail and bus companies are also technically public transport companies.

From the development of ferry services to cross rivers, public transport is virtually as old as cities, but many point to the *carrosses à cinq sols* (five-sol coaches) that operated in Paris between 1662 and 1677 as the first modern form of public transit. It was developed by the famous French scientist and philosopher Blaise Pascal. More sensibly, the advent of urban public transport systems started in the mid- to late nineteenth century, during the industrial revolution, similar to many other types of infrastructure like water and solid waste systems. After all, this is when cities started to grow in population and size. London itself went from having a population of around one million people in 1800 to three million people by the time the London Underground opened in 1863. This important population increase inevitably created an increase in travel demand. As we've seen, when demand increases, there are two options: control the demand or increase the supply. With such a significant population increase, strictly controlling the demand is impossible, and solutions need to be developed to increase the supply. As cities expanded into their hinterlands, new roads were built, and existing roads were widened, sometimes at the expense of entire neighborhoods.[1] London sought to increase its transport capacity by going underground. Coping with increasing travel demand and traffic congestion is not a recent problem.

---

1. As was the case in Paris during the time of Haussmann.

Public transport can be seen as a supply strategy to add capacity. Unlike physical infrastructure (e.g., roads), public transport also offers a mobility service. After all, people make trips, they travel from an origin to a destination, and the distance between an origin and a destination is often too great to walk or bike. Transit generally offers an affordable and convenient motorized service to make these trips—I do not want to get into the debate as to why transit is superior to private vehicles just yet, but it is; just wait a bit.

As the world population increased from 1.6 to 6 billion during the twentieth century—most often than not in cities—by the 2000s most large cities in the world had built extensive public transport systems. Beyond underground rail systems, transit systems can have many modes. The American Public Transportation Association (APTA) lists eighteen modes, the most common ones being paratransit, bus, bus rapid transit (BRT), light rail transit (LRT, including streetcars and tramways), heavy rail (including metro), regional rail, and ferryboat, whether powered by electricity or fossil fuels. Defining transit modes and differentiating between them is surprisingly challenging. We can try to do so based on technology (road versus rail versus water), capacity (e.g., number of riders per hour), right of way (separated versus shared with traffic), speed (slow versus fast), service provided (rush hour only versus all day), and other characteristics, and yet some transit experts keep debating.[2]

As a transit planner, the first consideration in selecting which transit mode to implement is capacity. If we turn this argument around, by first determining the expected demand (how many people will use the transit system?), we can then choose a service that meets this demand. For example, building a high-capacity system like a metro in a remote place does not make sense, and implementing a low-capacity system like a bus line on a high-demand route does not make sense either. Bus lines that run in mixed traffic carry between three and six thousand people per hour,

---

2. I know. I have been in such debates at the Transportation Research Board (TRB) Annual Meeting. It is more fun than it sounds, but it never ends with a consensus.

while metros carry between forty and seventy thousand people per hour—that is ten times more than buses. Except possibly for streetcars that run in mixed traffic and BRT systems, rail modes can usually carry more people, and they are more suitable to serve higher-demand corridors. Bus lines cannot carry as many people, but they are cheaper to implement and more flexible since they do not require rails.[3]

Some estimates of demand must therefore be computed to determine which mode is most appropriate. For this step, we must turn to the field of travel demand modeling, whose goal is to simulate travel demand and estimate how new systems and policies may affect demand. Travel demand modeling essentially tries to answer these four questions: Where do people come from, where do they go, which travel mode do they use, and which route do they take? Travel demand modeling involves sophisticated models that need to be able not only to simulate traffic in a city using some of the models discussed in the previous chapter but also to capture the behaviors of people in selecting a travel mode.[4]

After estimating demand and before selecting a mode, a second important consideration is the right of way (ROW). There are three types of ROW: fully separated, partially separated, and shared. The fully separated ROW applies to heavy (metro) and regional rail. It is often achieved underground or above ground, which makes it more complicated and expensive to build, but it is the best option to accommodate high-capacity modes, and it does not take away road space. The partially separated ROW applies to LRT, tram, and BRT. It requires physical separation from road traffic, except at intersections (whether barriers are used or not). It can accommodate medium-capacity modes, and it is much cheaper to build than the fully separated ROW (about ten times cheaper), but it takes away road space. Finally, the shared ROW applies to streetcars and

---

3. But they also do not leave that feeling of permanence that rail modes leave and that incentivizes the construction of new buildings that eventually results in short trip lengths.

4. The traditional travel demand modeling framework is called the urban transportation modeling system (UTMS), but it is better known as the four-step model because it is composed of four steps: trip generation, trip distribution, mode split, and trip assignment. Travel demand modeling is beyond the scope of this book, but it is fascinating.

buses. It does not require any physical separation, but it means that transit vehicles are stuck in traffic with private vehicles. It can accommodate only low-capacity modes.

There is a third factor that we should also consider: the long-term plan of the city. Transport is known to have an intricate relationship with land use—many say they are two sides of the same coin. Until now, we have assumed that all buildings are already built and that we are trying to satisfy an existing demand. Yet people move and new buildings are built, often influenced by the transport infrastructure around them. For example, new medium- and high-capacity transit services tend to attract new residences and new businesses. Cities therefore have the opportunity to incentivize where new developments should go. As a result, many cities have chosen to build systems that provide a higher capacity than needed to encourage investment in a certain part of the city. Some cities have even gone ahead and built new towns or neighborhoods altogether around rail stations. This is the case of Stockholm, which built satellite towns outside the city, all with railway stations with service to the T-Centralen station in the center of the city. This is also the case of Hong Kong, which famously developed entire neighborhoods around metro stations and sold them for profit.[5]

Finally, we should also consider the social landscape. Beyond the numbers, we should think about the people who need the transit service the most. While traffic engineers may see people as moving particles, transit planners consider the socioeconomic and demographic characteristics of transit users. For example, an effective transit service makes a big difference in neighborhoods where people may not have easy access to travel alternatives (e.g., for students, people with disabilities, lower-income neighborhoods). There are countless stories of people taking two hours to reach a destination, transferring three or four times to be able to

---

5. Instead of letting private developers collect the real-estate benefits from the new mobility service provided by the rail system (property values typically increase when new transit services are added), the transit company builds the neighborhoods itself. This is called *land value capture*.

go to work. Good transit planning offers an opportunity to make a real difference in the lives of millions of people around the world.

Now that we have estimated the demand for transit, identified the type of ROW that can be built, elaborated a long-term transport and land use plan for the city, and identified priority neighborhoods that should get transit service, we are able to select a transit mode and a preferred route. The next step is to select the number and location of stations.

There is a famous trade-off in transit planning when it comes to stations. Having many stations implies many stops, which translates into a longer journey (i.e., a long in-vehicle travel time). In contrast, having few stations results in a shorter journey, but the walk time to and from the station may now be excessively long (i.e., a long walk time). There is a sweet spot in the middle where the walk time plus the in-vehicle travel time results in the shortest journey. We will not go too deep, but that sweet spot can be found by summing all the separate travel times,[6] keeping the number of stations as a variable, and calculating the number of stations that results in the minimum travel time.[7]

There is one more type of "time" that we have not yet considered and that will be defined by the transit operator: it is the wait time—that is, the time spent waiting at a station for a vehicle to arrive. The wait time is related to the *frequency* of service—that is, how many times per hour is a station serviced (e.g., twelve times per hour)? The corollary is *headway*, which tells us the time between two consecutive vehicles (e.g., a bus every five minutes). From a user perspective, we all prefer a short wait time, but from the perspective of a transit operator, short wait times mean that more vehicles need to be dispatched that could have been dispatched somewhere else (plus running more vehicles is more expensive, and it does

---

6. In technical terms, the walk time is also called the *access* and *egress time*—egress is the time to walk from a station to a destination. The in-vehicle travel time is the sum of several travel times: acceleration time, cruise time, deceleration time, standing or dwell time (time for people to board/alight at a station), and intersection time (for partially separated and shared ROW modes). All of these times can be estimated for a transit line.

7. When we plot travel time versus the number of stations, we get some kind of a U shape, and the optimal number of stations is located at the bottom of this U. I detail this procedure in my textbook *Urban Engineering for Sustainability* (Cambridge, MA: MIT Press, 2019).

not necessarily make sense if nobody uses them or makes the most of their capacity). Finding the right frequency for a specific demand while making economic sense is challenging. At some point we have all cursed a transit company because we had to wait half an hour for a bus that was supposed to come every ten minutes. To be fair, I find that long wait times are not as big of a deal anymore, as many transit agencies make real-time bus locations and wait times available through smartphone apps.

Here, we have only discussed the planning of one single transit line, but planning and running an entire transit system is extremely challenging. There are too many variables to consider to be able to run an optimal system or find an optimal plan. Another variable is the number of transfers required to travel from any origin to any destination. Ideally, we want to minimize the number of transfers required, and therefore new lines can be added or existing lines can be modified to reduce that number. Of course, adding or modifying routes is effective only when transfer times are not unreasonable—that is another type of "time" that must be factored in: transfer time, another form of wait time. Again, we have all been there. Our bus arrives at our transfer stop, and as we get off, we see our transfer bus driving away, making us wait ten or twenty minutes or even more for the next bus. Careful transit scheduling is also essential to offer an effective service, but coming up with the right schedule to please everyone is practically impossible.

In large cities, transit agencies are gigantic enterprises, and transit planners have a monumental task. In 2024, Transport for London (TfL), the agency responsible for the entire transport system of London, had to manage 675 bus routes (with a fleet of nearly 10,000 buses) and 272 London Underground stations (running on 402 kilometers of rail lines) while being responsible for the Docklands Light Railway (DLR), the London Overground, the TfL Rail (the express service to the airport), the Elizabeth line, and the London Trams, all while managing a number of other assets and services as well. Overall, TfL handles around ten million journeys per day or roughly four billion journeys per year. These numbers are massive. And we have not even discussed the daily operation of transit systems involving bus depots, rail yards, maintenance issues, scheduling issues, management of breakdowns (in often aging and underfunded transit systems), and the hundreds or thousands of other aspects of running a transit system that are vital for a city. The phrase "Mind the Gap" in the title of this chapter refers to the beloved announcement familiar to anyone who has ridden the Tube: "Mind the gap between the train and the platform" (look it up, as you can easily find a recording online; the phrase even has its own Wikipedia page). When it comes to transit agencies, I want to say, "Mind the Complexity." Running a transit system is a complex undertaking. While I have a lot of respect for traffic engineers who handle all traffic signals in a city, transit planners and operators have my admiration.

So with all of the complications related to public transit, why is public transit superior to cars in cities? As we've learned, transit is a supply strategy to meet a demand. But what is demanded exactly? Sure, it is the service to travel from an origin to a destination, but if we look closer, we realize that what is actually demanded is an amount of space over time. By space, I mean the physical space, the actual area that you occupy as a traveler. By being packed in buses, trams, and trains, transit users simply take less space than people who travel by car, especially when cars carry only one person. That is how a city like London is able to have a higher population density than Los Angeles without collapsing. There is a famous picture showing the amount of space needed for sixty people to travel by car, bus, bike, and foot. Figure 7.1 shows an example from We Ride Australia. To me, this picture is one of the best examples of the idiom "an image is worth a thousand words."

Moreover, people taking transit do not need parking, whereas drivers do; that is again more space taken by cars. In Los Angeles, it should be clear now that the solution to reduce traffic congestion is not to build more roads but to build more transit and favor the right land-use policies to densify the city so people can walk to/from transit stops. On a similar note, I am always surprised when I hear drivers complaining that their tax money helps fund transit because they are not using transit. While they may not ride transit themselves, other people, who would have otherwise driven and made congestion worse, are. Even without mentioning the

Figure 7.1. *Road space by travel mode.*
Courtesy of We Ride Australia.

benefits to air quality, everyone gains from having more transit—people who ride transit and people who do not.

At this point, we have discussed private vehicles and transit, two modes that are great for long trips. For shorter trips, however, it makes much more sense to walk or bike. Walking and biking are a type of active transport (in contrast to private and public transport), and the city that perhaps best exemplifies active transport lies across the channel just east of London. I'm talking about one of the most famous biking cities in the world: Amsterdam.

BOGOTÁ'S TRANSMILENIO

London's transit system is world class.[8] That means it can be inspiring as well as intimidating. Many cities around the world are growing rapidly and would love to

---

8. Some elements of this story came from a discussion with Juan Acosta Sequeda, who is from Bogotá and was my PhD student at the time of this writing, and from the article "How One City Tried to Solve Gridlock for Us All" written by Michael Kimmelman and published in the *New York*

have a public transport system like London's, but it does not come cheap. Planning, building, and operating TfL's system is both expensive and resource intensive, two things that many growing cities in rapidly developing economies do not have. At least two strategies are available to cities aspiring to transform their transport system. Bogotá, Colombia, considered both in the late 1990s.

In the 1950s, around six hundred thousand people lived in Bogotá. To follow the path of modernization, the city hired the famous Swiss architect but terrible urban planner Le Corbusier to design a new masterplan for the city.[9] Anyone familiar with Le Corbusier will not be surprised to learn that the plan led to the removal of all existing train and streetcar lines and to the building of sprawling expressways. The new, revisited city was designed for 1.5 million people.

In the years that followed, during a period of intense internal conflict, many people fled the Colombian countryside, and many others moved to Colombia from unstable neighboring countries. By 1980, Bogotá's population soared to more than three million. By 2000, this number doubled to more than six million. By 2020, the population of the Colombian capital had reached eleven million.

What such a rapid, often low-income population growth requires is some form of transport service, but Bogotá did not have a public transport agency. When no formal transit service exists, the informal sector takes over. Informal transit service usually means having many small private companies offer semi-scheduled, unregulated, and unreliable travel services in vans and small buses that compete for passengers. In Bogotá, this episode is known as the *Penny War*, as service providers drove unsafely fast to pick up passengers.

As a city grows and becomes richer, transport inevitably becomes a major political issue. Thousands to millions of people rely on a well-operating transport system. By the 1990s, Bogotáns aspired to a better transport system, and they deserved one. Often, the preferred option is to look elsewhere, see what works, and reproduce something similar. That is the first strategy.

In the context of the 1990s, this first strategy led Bogotá to develop plans for an even larger network of expressways and a metro system that would make London proud. By now, we know why large expressway networks are seldom the solution—think of induced demand (expressways = more traffic congestion). What is less obvious is why a metro system like the London Tube is often the wrong solution for a growing city. Put simply, because they are underground, metro systems are terribly expensive to build. In fact, they are about ten times more expensive to build per kilometer than surface rail transit like LRT/tramway systems. Even if the

---

Times on December 7, 2023, available at https://www.nytimes.com/interactive/2023/12/07/headway /bogota-bus-system-transmilenio.html (accessed on May 29, 2024).

9. I have found that talented architects are often terrible urban planners. Le Corbusier is a great example. Another great example is Frank Lloyd Wright. The big difference between the two disciplines is that buildings are designed for individuals, whereas cities are designed for a society. While all cities are made up of buildings, cities are more than the sum of their buildings.

lines are elevated, building a metro system remains expensive as well as time con-suming. Although they are amazing and a great solution for cities that have the financial resources to build them (as we will see in Shanghai), opting to build a metro system usually means sacrificing coverage, thus servicing fewer people. Many cities in the world were lured by the London Tube and other world-class metro systems and have built one single line or a small network. In most cases, a surface rail line would have been cheaper to build, and the network could have been larger. That is why a second strategy is often preferable, a strategy we tend to overlook.

Namely, one should tailor a new infrastructure solution to your own particular needs. Forget about finding the best practice,[10] even if it seems to be the logical thing to do. Instead of copying what has been done in other cities, look at your problems and see how you can best address them with the resources that are at hand. We already had a flavor for it when we traveled to Hong Kong and learned about water distribution in Hanoi. This second strategy led Bogotá to build the TransMilenio, a large network of bus rapid transit (BRT) lines. BRT lines have a partially separated ROW, like LRT/tramway systems, but they run on rubber wheels like buses. They are significantly cheaper to build than metro and LRT/tramway systems, which means that coverage can be increased. Moreover, they can be built more quickly, which means that people get access to an adequate transit service rapidly. BRT buses are also large and articulated and can accommodate many people (unlike buses that run in mixed traffic). The TransMilenio started operating in December 2000 and expanded to 114 kilometers (71 miles). It quickly serviced millions of people daily. That is more than most metro systems in the world.

The TransMilenio became a major success and a model for many cities world-wide. However, when Enrique Peñalosa, the mayor who started it, left office, the service deteriorated. Other parties did not support the system as much. The service coverage was supposed to be expanded to 387 kilometers (241 miles), but it never was. Millions of dollars of funding for the TransMilenio were instead embezzled. The bus fleet was poorly maintained. Peñalosa was elected again in the 2010s and revitalized the service and even added another transit mode, this time a gondola lift service called the TransMiCable to link predominantly lower-income populations in hilly areas to jobs and schools.

As of this writing, Bogotáns still aspire for more and better transit. After much pressure, Peñalosa supported the construction of Bogotá's first metro line, to be elevated (as it is cheaper to build). Plans started in 2016 for a scheduled completion in 2028. Line 1 will be twenty-four kilometers long (fifteen miles) and serve

---

10. For complex systems such as cities, copying examples of best practice is not necessarily a good idea. See Dave Snowden's Cynefin framework. It was developed for organization management, but I think it applies to infrastructure as well.

sixteen stations. That is a long time for a small coverage. While I am a fan of metros, I must admit that BRT systems are often preferable, especially when resources are tight, to get a system built fast to serve people who often desperately need transit service. Maybe a major expansion of the TransMilenio would have been preferable here again. Despite the coolness of a metro, people need transport solutions quickly and at a reasonable cost.

Another solution that Bogotá implemented was building a large network of bike paths. By 2023, the system was close to 600 kilometers long (372 miles), and there are plans to expand it. From 1 percent in 1996, cycling now makes up 17 percent of the travel mode share, but we are getting ahead of ourselves. Active transport is the topic of the next chapter.

In the end, we all have a lot to learn from the Bogotá story. Let us not forget the second solution: tailor a new infrastructure solution to your needs. I think it applies to all infrastructure systems, not just public transport.

CHAPTER EIGHT

Active Transport

Pathways, Bikeways, and Shared Spaces in Amsterdam

ONE OF MY FAVORITE BOOKS OF ALL TIME IS *THE FALL* BY ALBERT
Camus. The story takes place in Amsterdam. The protagonist is Jean-
Baptiste Clamence, a "judge-penitent," who spends his days in a bar
named *Mexico-City*.[1] Camus gives a grim account of the city. From the
book, I pictured it as bleak, unwelcoming, and constantly gray. Camus
even makes an analogy between the concentric canals of Amsterdam and
the circles of Dante's *Inferno*—it is hard to go bleaker than Dante's
Inferno. My expectations were therefore low when I got the chance to visit
the city during a four-day January trip, although I must admit I was
excited to see the canals, to ride the tram, and to visit the city's museums.
But how wrong I was! Instead of bleak, I found it beautiful and vibrant
with tons of street life. I found a city full of culture and tasty food—you
should taste *stroopwafel* and raw herring in Amsterdam.[2] Most vividly,
I remember having quaint walks across the city center, especially in
the Grachtengordel-West neighborhood, and I remember the constant
zooming of bikes and the immense bike parking lot by the main train
station. Really, what I found was a haven for active transport.

---

1. This bar has never existed in reality. I know—I looked hard for it.
2. For breakfast, the Dutch also cover their toast with chocolate sprinkles, the kind of sprinkles
you normally put on soft-serve ice cream.

Active transport is a means of transport that uses human physical activity as the primary motor. The two main active transport modes are walking and biking, but any means of transport that require physical activity like running, skateboarding, roller-skating, scooter riding, and using a wheelchair are included as modes of active transport. By far, active transport is the preferred travel mode when available. Being affordable, healthy, and environmentally friendly, active transport has got it all. It includes the oldest transport mode: walking. Yet the science and engineering around active transport are surprisingly not well developed. To my knowledge, the aspect of active transport that is most studied is *safety*. But even in this area, safety relates to the interaction between active transport and other transport modes, as opposed to being contained within active transport. This chapter deceptively contains less engineering material because, as of this writing, transport engineers do not tend to deal much with active transport. My guess is that we have taken active transport for granted because it had been the main travel mode forever. In most cities, however, active transport is not the main travel mode anymore.

We've already discussed travel demand and the capacity to meet a demand. We must now discuss the concept of mode share. *Mode share*[3] measures how much each travel mode is used. Usually, mode share values are reported as a proportion in the number of trips—that is, the percentage of trips made using private vehicles (cars, SUVs, etc.), riding transit, walking, or biking, and so forth. Mode share values vary greatly across cities. In the L.A. region, the mode share distribution is as follows: 89 percent private vehicles, 5 percent transit, 3 percent walk, and 1 percent bike. In Greater London, it is 37 percent private vehicles, 35 percent transit, 25 percent walk, and 3 percent bike. And finally, in the Amsterdam region, it is 42 percent private vehicles, 19 percent transit, 5 percent walk, and 30 percent bike.[4] How different!

---

3. Or *mode split* or *modal share* or *modal split*.

4. Note that we must be a bit careful when comparing mode share values. We need to make sure that we are comparing the same things. What geography are we using (inner city versus metropolitan

From these values, we can see how building infrastructure influences demand. L.A. has a strong history of building roads, and, as a result, most people drive. London has a great transit system, and, as a result, it is used heavily. Amsterdam has heavily invested in its active transport infrastructure, and, as a result, most people bike or walk. The mere presence of infrastructure makes a big difference in the long term, both on transport and on land use. This is why it is time to advance the science and engineering of active transport infrastructure.

We can start by inventorying the different types of infrastructure for active transport, focusing on walking and biking. This is an exercise that the U.S. Transportation Research Board (TRB) has done in its famous *Highway Capacity Manual* (HCM), which, notably, details the level of service (LOS) that we learned about in L.A. Despite its name, the HCM does not focus solely on highways. It also covers public and active transport—but the title is indicative of the predominance of the private mode in the United States.

For walking, the HCM reports ten different infrastructure types: walkways, sidewalks, plazas, pedestrian-only zones, staircases, queuing areas, road crossings, underpasses, overpasses, and shared pedestrian-bike paths. To make walking possible, infrastructure needs to be built for people to walk from their origin to their destination. This is why the presence of sidewalks is particularly important, albeit not always available. The presence of crossings is also important, and, for walking, the presence of more crossings is preferred. I have walked on too many roads, especially in business and commercial centers in suburban neighborhoods, that have large sidewalks (thus seemingly inviting to pedestrians) but where just reaching an intersection with a crossing takes a long walk. At the intersection, it takes a long

area)? How is mode share calculated (trip based versus distance based)? What mode share values are we reporting (commute-only trip versus all trips)? What modes are included (note that the mode share values I report do not sum to 100 percent for L.A. and Amsterdam; the remaining trips are taken using other modes like taxis/ridesharing). Here, I cite data from the Deloitte City Mobility Index 2020 that reports the geographical boundary selected (L.A. is the LA–Long Beach–Anaheim area, London is Greater London, and Amsterdam is the Amsterdam Transport Area), but I could not find information about the dates when the data were collected. Mode share is also called "journey modal split," but I am not sure what "journey" means here (i.e., commute trip versus all trips).

time for the walk signal to come on, and when it does, the duration is so short that I see the red hand waving at me when I am barely halfway across. Regardless of how much one wants to walk, these roads are not made for pedestrians. The right infrastructure is simply not there.

In contrast, I have walked in many neighborhoods, including many in Amsterdam, with adequately sized sidewalks, short blocks that allow easy crossing, sometimes with pedestrian paths and small plazas that are pleasant to cross (and sometimes equipped with green infrastructure). Often, these neighborhoods also have some street activity that makes one feel safe. This is what Jane Jacobs,[5] arguably the best thinker of urban planning and design in modern history, calls "eyes on the street." This is also partly why she favors medium-rise buildings with shops on the first floor. The butcher, the hairdresser, the baker, and other shop owners and employees provide "eyes on the street," making it more inviting to walk. In the end, with the right infrastructure, the experience as a pedestrian is completely different.

For biking, the HCM reports six types of bikeways. Four of them are on-street options: shared lanes (mixed with cars), bike lanes, shoulder bikeways, and bike tracks; the other two are off-street categories: side paths and exclusive pathways. The mere presence of a dedicated bikeway—not a lane shared with other vehicles—is known to make a significant difference for bike use. In particular, dedicated bikeways significantly improve traffic safety, although dooring (collision between a bike and someone opening the door of a private vehicle) is a significant problem if a bikeway is not properly designed. And what is a properly designed bikeway? One of the best guides I have seen does not come from a reputed engineering firm[6] but from the National Association of City Transportation Officials (NACTO), a U.S.-based association of cities and transit

---

5. When it comes to cities and urban planning, Jane Jacobs deserves an entire chapter to herself. I even wrote extensively about her and her lessons in my textbook *Urban Engineering for Sustainability* (Cambridge, MA: MIT Press, 2019).

6. Which, to me, demonstrates how many transport engineers are dropping the ball on their responsibility to properly design transport infrastructure as a whole, at least in the United States, but I know it is also the case in many countries around the world.

agencies. The American Association of State Highway and Transportation Officials (AASHTO) also has published guides for both bicycle and pedestrian facilities.

To bikeways, we can add bike parking as an important type of bike infrastructure, from individual bike racks on sidewalks (e.g., inverted U-shaped racks locked in the ground) to larger multibike racks off sidewalks, in plazas, and in front of buildings, all the way to large bike parking facilities that can accommodate hundreds of bikes such as the one by Amsterdam Centraal (Amsterdam's central train station). Having dedicated bike parking infrastructure is essential to make it inviting for people to bike—walking around Amsterdam, you can see bike parking facilities on every street in addition to railings and fences that are also used to park bikes. Finally, there is another type of bike infrastructure that became more popular in the early 2000s: bike-share systems. From Divvy in Chicago, Ecobici in Buenos Aires, and Vélib' in Paris to PubliBike across Switzerland, YouBike in Taiwan, Mobike in various cities around the world (including Berlin and Beijing), and the list goes on,[7] bike-share systems have sprouted in many cities around the world. Some of these systems have docks, where a bike can be picked up and where it needs to be returned; others are dockless and use an app to unlock/lock a bike. As of this writing, Amsterdam has at least four popular bike-share programs: Donkey Republic and FlickBike (dockless bike sharing), Urbee (electric bike sharing), and Cargoroo (bikes with a large storage space in front to carry children or goods). These bike-share programs are sometimes publicly owned and sometimes privately owned. They tend to enjoy mixed success, doing well in some cities and not so well in others. Their success partly relies on their service offering. In many cities, bike-share programs compete with public transport, costing roughly the same for a ride, while distances that can be traveled are more limited than with transit. This is often not the right strategy, since both modes should be complementary

---

7. PBSC reported that in 2024 there were over two thousand bike-share systems in the world and over nine million shared bikes; see https://bikesharingworldmap.com/ (accessed May 29, 2024).

as opposed to competing. That's part of an integrated transport strategy (more on that topic soon).

After infrastructure, the second element to study for walking and biking is demand. Unlike traffic and transit, active transport generally does not suffer from congestion. I use the term "generally" because there are exceptions. Both bikeways and sidewalks can be extremely busy, especially during rush hour. In transport, there is a phenomenon that applies to all modes called *platooning* when surges of people arrive at the same time.

Traffic congestion can therefore apply to active transport as well. Escalators offer a funny but illustrative example. The social convention is that if you prefer not to walk up/down an escalator, you stick to one side. Half of an escalator therefore has people who are standing and the other half has people who choose to walk up/down. That seems logical, right? This way everyone is happy. If you want to walk up/down, you do, and if you don't want to walk, you don't. But when it comes to traffic, do you think this is the right strategy to maximize the flow? When it gets busy (e.g., when a metro arrives at a station), many people end up queueing at the bottom of the escalator—a typical example of platooning. So, what is the right strategy? Although it first appears counterintuitive, the right strategy is to forbid people to walk altogether. Essentially, when people can walk, half of the escalator is relatively empty. In contrast, by forbidding people to walk, the full capacity of the escalator can be utilized. At a metro station in London, planners found that flow was increased by 27 percent when people were forbidden to walk up/down. Slower speeds resulted in higher flows.[8]

Looking into our traffic engineering toolkit, it does not take long to realize that we can even calculate the LOS of walkways and bikeways as well, which you can find in the HCM. The concepts used are identical to the ones developed for traffic engineering; the units of analysis are now average meetings or passings per minute, path width, or other variables

8. I read about this story in the article "Escalator Etiquette: Should I Stand or Walk for an Efficient Ride?" published in *The Conversation*, available at https://theconversation.com/escalator-etiquette-should-i-stand-or-walk-for-an-efficient-ride-112287 (accessed May 29, 2024).

originally defined to be used for vehicular traffic. In fact, I would not be surprised if the LOS methodology applied to active transport was developed by traffic engineers (the HCM also details a methodology to calculate the LOS of transit). Because the LOS is calculated based on travel speed, it inherently emphasizes travel time, but in active transport, comfort, enjoyment, and safety are often more important. In his 2015 TED talk, the researcher Daniele Quercia explains how he used to take a route with his bike because his smartphone app recommended it (i.e., it was the fastest route). Months later, he found an alternative route himself that would take a few minutes more but avoided heavy car traffic, felt safer, was more enjoyable, and was more comfortable. From this experience, he started the concept of Happy Maps in which street segments were scored based on how "happy" they made people feel by analyzing the text[9] from geolocated social media posts. Happy Maps can then suggest paths that do not significantly increase travel time (up to 10 percent longer than the fastest path) but can significantly increase happiness. In active transport, should minimizing travel time or maximizing flow be the main goal? Or should we focus on providing an enjoyable, comfortable, and safe experience to people walking and biking?

The HCM manual offers amazing guidelines, especially when it comes to private vehicles, but it is time to stop focusing on flow optimization, especially in active transport. The analogy with stormwater management can be made again. Let us not try to "get rid" of people, like we used to get rid of stormwater as quickly as possible by maximizing flow; instead, let us think about the entire experience. To my knowledge, most transport engineers are trained to focus on flow optimization and ignore this notion of experience (some emphasize air quality). This tendency is less true in non-engineering disciplines such as urban planning and urban design that emphasize experience. In fact, there is one approach in urban design from which transport engineers have much to learn. It is that of *shared space*.

---

9. This is a method called *sentiment analysis* in natural language processing.

Shared space promotes the idea that streets should be shared across all modes. The central idea of the approach is the removal of all barriers between modes: no sidewalks, no bikeways, no crossings, no road surface markings, and so on. Often, barriers were put in place at first for the sake of safety. But separating modes has led to an increase in traffic speed, which has in turn led to a decrease in safety. What we are left with are segregated corridors of fast traffic that offer no comfortable experience. Instead of harboring the vitality of cities, streets have become transient places to go from one building to another. It is like we are told to avoid spending time outside; outside is not safe, regardless of how many barriers we install. But look at the most attractive areas in cities, the places to visit—usually they will be streets with shops and activity. I am always wary when I am told that the "place to see" in a city is the mall. Personally, I remember going to Atlanta in 2010—another car-centric city like L.A.—and asking what I should do. I was recommended to go to an outlet mall like there are in cities across the United States. When I went again in 2019, I was told to go on West Peachtree Street NW that had transformed and become active. What a different experience. I hope that next time I visit Atlanta, West Peachtree Street NW will have become a shared space.

Furthermore, shared spaces often have pavers instead of asphalt, which can potentially be turned into green infrastructure as well.[10] The use of pavers helps reduce the speed of cars, which can help increase the flow of cars—back to our fundamental flow diagram. No markings are painted on roads, so drivers, bikers, and pedestrians pay more attention. In Amsterdam, like in many European cities, many streets exhibit features of shared spaces, often because they are old and were built as shared spaces in the first place. It might therefore not come as a surprise that the idea of

---

10. Although we should be a bit careful since the science and engineering around green infrastructure design are not perfect yet. For example, we should account for the load from heavy vehicular traffic that impacts the subgrade that contains the void that stores stormwater. Green infrastructure also requires maintenance that would impact traffic. At the time of this writing, green infrastructure using pavers was preferred in parking lots, walkways, and bikeways, but I hope their use will be extended to streets soon.

shared space came from a Dutch person. Hans Monderman was a traffic engineer and road safety specialist, but he defied the philosophy of his peers that sought to segregate modes to improve traffic safety and flow; so far, the projects that have been built are proving him right. A quote of his particularly encapsulates his philosophy: "Most engineers, when faced with a problem, try to add something. My instinct has always been to take something away."[11]

The science and engineering around active transport are therefore still at their infancy, but we can see how seemingly silent infrastructure such as sidewalks, intersections, and road space markings can have such an influence on people. Many cities around the world have started to transform themselves to promote active transport to increase street life, which is known to increase the livability and economy of a city as well. I would not be surprised if cities around the world completely transformed themselves. While they have expanded substantially into their hinterlands in the twentieth century, cities will transform themselves *from within* in the twenty-first century. While the private vehicle shaped the cities of the twentieth century, it is the integration of all modes that will shape the cities of the twenty-first century. In fact, we need to push this idea further. That point brings us to the notion of integrated transport. For that, we travel east. It is time to visit Shanghai.

PHILOSOPHIZING IN KYOTO

It is hard to overstate the importance of active transport. Humans have always traveled, and for the longest time, the only travel mode available to humans was walking. Beyond its most basic function of traveling, walking also possesses many benefits. Aside from its obvious health benefits, walking offers a means for people to think. According to the legend, the Greek philosopher Aristotle used to walk while lecturing, which led him to found the Peripatetic school (from *peripatētikós*,

---

11. The quote is similar to one of my favorite quotes from another one of my favorite writers, Antoine de Saint-Exupéry, who wrote in *Winds, Sand and Stars*, "A designer knows when they have achieved perfection not when there is nothing left to add, but when there is nothing left to take away."

meaning "of walking"). In the early twentieth century, two Japanese philosophers enjoyed walking and chatting along a path near a small river in Kyoto. Today, that path is known as *Tetsugaku-no-michi*, or (in English) Philosopher's Walk.

Philosopher's Walk has become a popular visiting spot in Kyoto, especially in the spring when the cherry trees are blooming. The walk is quaint. Many people enjoy taking it after visiting the Ginkakuji temple (the silver pavilion). The small stream along the walk is soothing and perfect for engaging in a philosophical discussion—do you remember Heraclitus's aphorism "panta rhei" (everything flows) from the chapter on water collection in Rome?

The absence of heavy car traffic by the stream is particularly enjoyable. Several temples along the way can be visited. Kyoto must be one of the most culturally rich cities in the world. There are temples and cultural venues all throughout the city. Kyoto used to be the capital of Japan; it is where the emperor lived until the late nineteenth century. In fact, the name Kyoto itself translates to "capital city," while Tokyo translates to "eastern capital."

Despite the millions of tourists who visit the city every year, the many small streets enable everyone to develop an intimate relationship with the city. It is quite a feat. In particular, many streets are fairly narrow and inconducive to car traffic, which leaves ample and safe space for pedestrians and cyclists. Note that car traffic is not prohibited in these streets; their small design is just not conducive to driving. To me, these narrow streets offer a perfect example of shared space. Everyone must pay attention, whether on foot, on a bike, or in a car.

The blocks are also relatively short, around 140 meters, making it easy to walk around the city, to reach your destination quickly, or sometimes to get lost happily while wandering around and finding an amazing pastry shop.[12] In contrast, I remember complaining when I lived in Toronto. The blocks are around 320 meters long, and going from one end to the other was annoyingly long, especially when walking in the winter or along heavily trafficked streets.

I have visited Kyoto three times, each time with different people, and each time we all left impressed. Although I never biked around the city, many people did. I could easily see myself biking around the city as well and feeling safe. The transit system was also convenient and easy to use thanks to my smartphone.

Older cities like Kyoto have a significant advantage over newer cities like Toronto. They were built at a time when cars did not exist. They were built for pedestrians. Newer cities like Toronto are car-centric because they were built when cars became popular. Active transport tends to be less enjoyable in cities that grew in the car era because they were not built for people but for cars. The infrastructure that makes active transport so great in older cities is absent in newer

---

12. My wife fell in love with a small, high-end, Zen pastry shop not far from the Yasaka-jinja Shrine.

cities. It is high time to advance the science and engineering of active transport to help every city, old and new, offer an experience as enjoyable as Kyoto does.

At the end of Philosopher's Walk, you can walk down the hill, visit the Heian-jingū Shrine, and keep walking until you find the canal that brings the water from the stream along Philosopher's Walk to the Kamo River. Along the way, you will find another marvel. It is not a cultural marvel this time but a small engineering marvel: the Ebisugawa Dam, a small-scale hydroelectric power plant. I do not think I would have ever found the Ebisugawa Dam if I had visited the city by car or transit. There is something about active transport that fills the air with magic.

Chapter Nine

Integrated Transport

The Only Way to Prevent Transport Chaos in Shanghai

WHEN I LANDED AT THE SHANGHAI PUDONG INTERNATIONAL AIRPORT in 2016 to attend a conference, I was excited. After going through customs and immigration, I entered the magic doors that open to the arrival hall with the usual crowd of people waiting for passengers. As a real transport nerd, I frantically followed all the signs that featured an odd-looking symbol, half-train, half-crab, with the letters "maglev." Traveling at 430 kilometers per hour, the maglev—short for magnetic levitation—is a rail mode that uses magnetic forces to make railcars float in the air as opposed to resting on rails.[1] The Shanghai maglev gets you from the airport to Longyang Road Station in eight minutes compared to half an hour by car. It is a great experience. But the story is not over yet. To go downtown, then, you need to take the metro. Since opening in 1993 (130 years after the London Underground), the Shanghai metro has become the largest metro system in the world. Its construction is a feat, but a feat that was badly needed.

The population of Shanghai soared from six million people in 1980 to nearly thirty million in 2020. New districts were built in record time. In

---

1. Benefits of the maglev technology include the fact that there are no moving parts and no friction with rails. But maglev is also much more expensive, and, as of this writing, the benefits compared to more conventional high-speed train technology did not justify the extra costs. When I returned to Shanghai in 2023, the maglev had stopped operating.

the thirty-year span from 1990 to 2020, the city became known as the fastest-growing city in the world.[2] Massive infrastructure was built for the 2010 World Exposition, but the most impressive construction happened in the Pudong New Area that houses the Shanghai financial district, proudly featuring the iconic Oriental Pearl Tower.[3] And this growth is not over. When I traveled to Shanghai, I visited the Shanghai Urban Planning Exhibition Center, and I saw that the city will keep transforming itself. I also saw how the city has managed to make the "old" cohabit with the "new." I learned about Shikumen architecture, which evolved in the second half of the nineteenth century and blends Chinese and Western styles. I visited the Former French Concession area that is known for its historical and charming architecture. And from a culinary perspective, I visited many delicious xiao long bao joints—a must for everyone visiting Shanghai.

Accommodating an "old" city for a population increase of twenty-five million people in less than thirty years is a nightmare for transport planners. Beyond the metro system, the only way to avoid transport chaos with thirty million people traveling across a region is with integrated transport.

So far, we have discussed transport modes as competing with one another: private vehicles versus transit versus walking and cycling. Even at the personal level, who has not complained about people using other modes than the one we are using? As a pedestrian or cyclist, I find that some cars drive too fast and present a hazard. As a driver, I find that cyclists should be more careful in mixed traffic. As a transit rider, I despise when a driver swerves right in front of my bus. After all, humans are territorial animals, and we quickly and easily develop visceral feelings when

---

2. I am not sure what metric was measured, but it could be population. A 2020 article from the International Institute for Environment and Development showed that Shanghai was then the second fastest growing in the world in terms of population, second to Delhi (India).

3. The short article "26 Years of Growth: Shanghai Then and Now," published in *The Atlantic* (2013), shows a picture of the Pudong New Area in 1987 and in 2013. The contrast in the two pictures looks surreal, out of a science fiction movie: https://www.theatlantic.com/photo/2013/08/26-years-of-growth-shanghai-then-and-now/100569/ (accessed May 29, 2024).

someone outside of our territory (like a person using another travel mode) attacks our territory.

In the past three chapters, I have made an argument for a hierarchy of modes. First, active transport is the preferred travel mode; people walking and cycling are the livelihood of cities. For longer distances, a motorized transport mode is needed; in this category, public transport is preferred since transit riders take less physical space. At the bottom of the hierarchy come private vehicles. Many urban thinkers vilify private vehicles. Notwithstanding problems related to air quality and greenhouse gas emissions,[4] the dominance of private vehicles has figuratively killed many cities by killing their livelihood: their streets. In many cities, streets that used to be bustling with people, giving life to a city and offering eyes on the streets, have become empty. A movement for the "slow city" is growing in the world, in which active transport is prioritized and strategies are adopted to slow down traffic. It might sound counterintuitive, but slower traffic translates into a panacea of benefits, including smoother traffic, more livable and economically thriving cities, and fewer greenhouse gas emissions. It is a recipe for success.[5]

Jane Jacobs, whom we briefly met in the last chapter, wrote a book titled *The Death and Life of Great American Cities* that was published in 1961. In the book, Jacobs largely puts the blame of the death of cities on cars. In a diverse world, too much of one thing is bad—I do not like generalizations, but I often find myself saying that "diversity is always good." The truth is that we live in a complex world, and it is by nurturing this complexity—notably through nurturing diversity[6]—that we can make it

---

4. That problem would partly go away with electric vehicles.
5. You can learn more about this idea in various works, including P. Tranter and R. Tolley, *Slow Cities: Conquering Our Speed Addiction for Health and Sustainability* (Amsterdam: Elsevier, 2020). I also personally contributed to the subject in J. Mulrow and S. Derrible, "Is Slower More Sustainable? The Role of Speed in Achieving Environmental Goals," *Sustainable Cities and Society* 57 (2020): 102030.
6. In all its meanings, not just racial or gender diversity. Diversity means different types of businesses, land uses, food options, and so on, and of course racial, ethnic, and gender diversity, as well as travel options.

livable.[7] This statement directly applies to transport. By focusing on one single travel mode, we are killing part of the diversity. Another great urban thinker, Lewis Mumford (whom we met in the chapter on traffic and roads in L.A.), wrote in his 1961 book *The City in History*, "Currently the most popular and effective means of destroying a city is the introduction of multiple-lane expressways, especially elevated ones, into the central core. This came about immediately after elevated railways for passenger service were being demolished." (Mumford also points to the preceding undesirable dominance of railways.[8]) And after decades, many cities went about destroying their elevated expressways.

As the comedian Jerry Seinfeld said in his 2020 show *23 Hours to Kill*, "Nobody wants to be anywhere." We are constantly hurrying to get somewhere, and when we are there, we are thinking about how we are going to get back. Instead of vilifying cars or picking sides and promising obedience to one single mode, we need a balance between modes, which cannot happen without integrated transport.

Let us go back to the basics: Why do we travel? To go from an origin to a destination. In most cases, we do not care which travel mode we take as long as the journey is as short, as affordable, and as comfortable as possible. That is why people in L.A. drive (even for short trips less than a kilometer long), why people in London take transit, and why people in Amsterdam bike. As far as I know, Angelenos, Londoners, and Amsterdammers have no special genetic or physical attributes that make them prefer to drive, ride transit, or bike; they just seek to travel in a way that minimizes the time spent traveling and maximizes affordability and comfort. Now, what if instead of using one single mode to travel from the origin to the destination, you could use multiple options? What if you

---

7. For the fans of *entropy* out there, in society, most times, we should seek to be in a high entropy state so that no one element dominates.

8. In fact, from streetcars to regional rail lines, the initial driver of suburbanization was the rail mode, not the car. You can read more about this subject in K. H. Schaeffer and E. Sclar, *Access for All* (New York: Columbia University Press, 1980). The book offers an insightful history of the relationship between transport modes and urban growth.

could drive a short distance to a train station, take the train, exit a station, and walk or bike to your final destination? Here, we are using the three main travel modes used in L.A., London, and Amsterdam.

To make this combination of travel modes possible, they need to be integrated. In Shanghai, regardless of how large the metro network is, it cannot serve all thirty million people by itself, but it can be part of the journey for many if it is integrated with other travel modes. This is why many train stations in suburban areas have parking facilities and bus lines that bring users to them—they are called *feeder buses*. This is also why many bike-share systems have docks close to train stations. These are signs of integrated transport.

There are at least three types of integration in transport: physical, operational, and fare integration.

As the name suggests, physical integration is about integrating different transport modes together physically. Transit offers a fitting case, as many bus lines stop by train stations to integrate the two transit modes: buses and trains. Having parking facilities and bike-share docks at a train station is also an example of physical integration, and so was the story in the previous chapter in Amsterdam in which an enormous bike parking facility was built next to the main train station. Integration here is not obvious because of land ownership. For example, space needs to be available to build parking lots around train stations. Local governments need to approve the installation of bike-share system docks on city property. Although not between two urban travel modes, another famous type of physical integration is between airports and train systems, as was done at the Shanghai Pudong International Airport. Taking a train from an airport to a city is generally more comfortable than taking a bus, but, at times, train stations are built far from airports, resulting in excessively long walks to the station (which are painful after long flights).

To go one step further, one popular type of physical integration that has been transforming cities all over the world is encapsulated in the term "complete street." Complete streets are a relatively new way of designing streets. The principle is that streets are made for everyone, and therefore they should be designed for everyone. This idea means that streets should offer a pleasant experience to drivers and transit riders and cyclists and pedestrians and everyone else traveling, regardless of their physical ability. Complete streets often have bike paths and are served by some type of transit. Walking signals to cross the street are timed so that everyone has ample time to cross them. Drivers feel comfortable driving through a complete street. Complete streets also often boast businesses. They are pleasant to walk along and have many "eyes on the street" so that people feel safe. Shanghai has been at the forefront of retrofitting its streets to fit this model, converting many of them from vehicular traffic corridors to complete streets. Complete streets are different from shared spaces that we learned about in the previous chapter in Amsterdam, but we can see that some of the goals are similar: streets are for everyone.

We can extend physical integration another step further with transit-oriented development (TOD). TOD is a type of physical integration that revolves around properly developing and using the land around transit hubs to make it conducive to use transit—not just streets, all the land. TOD designs range from building adequate pedestrian infrastructure to make it easy to walk to a station to adopting the right land-use policies by building residences and businesses within a short walk of a station to favor the use of transit (as opposed to private vehicles). Some TOD designs also directly integrate building complexes with transport hubs so that one does not even have to go outside to reach a shopping or a residential complex. Stockholm and Hong Kong offer good examples of TOD designs, as we saw in the public transport chapter. Many cities around the world increasingly try to develop policies and solutions that integrate transport and land use, including in Shanghai. Physical integration is challenging but an essential first step toward successful transport integration.

The second type of integration is operational integration. Operational integration is especially important between service providers—for example, coordinating routes so that the same service is not offered by multiple providers while other routes that would be needed are ignored. Another goal is to coordinate schedules to minimize wait time, especially for low-frequency lines (so that your bus or train does not leave just as you arrive at a stop, making you wait twenty minutes for the next service). From a technical perspective, coordinating routes and schedules can be a nightmare for large, complex systems, such as in Shanghai. How can you ensure that a bus in mixed traffic arrives at a station a few minutes before the departure of a train? To make matters more difficult, many large cities have separate private operators running transit lines, who may or may not choose to cooperate.[9] In many cities, regional authorities are created simply to coordinate all travel modes and ensure the best service possible for the people of the region. This is the case in London, for example, where Transport for London (TfL) awards routes and schedules to private operators who are rewarded or penalized depending on their performance. The establishment of these authorities is vital for the successful integration of transport systems. In fact, the mandate of these authorities generally goes beyond transit. As we saw in the public transport chapter, TfL is responsible not only for the London public transport system but also for the entire transport system as a whole, including all roads, bridges, and even water transport (the Thames is an integral part of the London transport system).

The last type of integration is fare integration. In order to truly integrate transport modes, the payment method should be seamless. This point is especially important for public transport. In many cities, individual transit lines were developed by private entities that developed their own payment strategy to capture and retain their customers (for example, having different token sizes so you could not use the same fare payment mechanism across systems). People transferring between systems had to get another pass/token, oftentimes having to buy a new fare altogether. This practice was always a hassle and a deterrent to using public transport. Transit operators sometimes did not care, as they enjoyed a monopoly or,

---

9. Funnily, in large cities where demand for transit is significantly high, trains run so often (e.g., every five minutes) that private companies do not need to coordinate schedules, as is the case in Tokyo. This is a good example of how chaos can make some things easier.

in contrast, they wanted users to pay a second fare since the company was not doing well financially (running a transit company is tricky). Over the course of the twentieth century, and as cars became increasingly popular, many transit companies went bankrupt and were taken over by local governments. This is partly why most transit systems now are either government owned or privately owned but government funded (i.e., they do not collect fares; the government does and pays them to have a more stable income). Fare integration remained a problem for a long time. Everything changed in the early twenty-first century, however, thanks to the arrival of internet and communication technologies. In most large cities, through the use of an app or a smart card, it has become easy to seamlessly transfer between systems without even knowing that the systems are different. From the Ventra card in Chicago, the Oyster card in London, the Octopus card in Hong Kong, the Pasmo in Tokyo, and the Myki in Melbourne, smart cards have often become the face of public transport systems,[10] and some transport nerds (like me) enjoy collecting them. My personal favorite is the EZ-Link Card in Singapore that can be used not only in the public transport system but also in taxis, to pay road tolls and parking, at gas stations, and even in many stores throughout the city-state; this is true fare integration. Similarly, the Shanghai Public Transport Card—the Jiaotong Yikatong in Chinese—can be used to pay for a suite of transport services, including transit, taxis, parking, fuel stations, tolls, and, of course, the maglev.

In Shanghai, the Shanghai Municipal Commission of Transport is tasked with looking after all transport-related matters. The commission was given sixteen organizational functions that extend well beyond the three types of integration discussed earlier. While the first function focuses on the implementation of transport laws, regulations, rules, and policies, the second one develops "comprehensive transportation development strategies . . . optimizes the layout of transportation structures,

---

10. As of this writing, Wikipedia listed more than seven hundred public transport smart cards around the world; see https://en.wikipedia.org/wiki/List_of_smart_cards (accessed May 29, 2024).

coordinates various transportation modes, and promotes multimodal transport development."[11] The famous Spider-Man quote "with great power comes great responsibility" should be turned around when it comes to integrated transport. It should be "great responsibility requires great power." In some cities or countries, transport integration plans can fail because regional authorities are given the responsibilities on paper but not the power to change anything, but great power is needed as cities around the world are transforming how their residents travel.

In the first half of the twenty-first century, a new way of thinking about urban transport emerged. Beyond competition between travel modes and beyond smart cards that may soon be obsolete, urban transport is being rethought around one concept: Mobility as a Service (MaaS). MaaS revolves around the idea that owning a personal vehicle has become impractical and expensive for many urban residents. Instead, it becomes easier, more affordable, and more comfortable to use a service instead. From riding transit to borrowing a bike or an e-scooter, renting a car for short durations, or ridehailing/ridesharing a car, many mobility services are offered to city residents that can replace the traditional model of committing to one single mode with using whichever mode is best at a given time. The idea is to have people choose mobility services like they pick other services, such as mobile phone plans or internet services. What type of mobility service would you like? How about a monthly package with forty transit rides, three hours of car rental or taxi, and two hours of bike share? You can pay anything extra by the ride or by the number of minutes you use a car/taxi. MaaS is possible only when all transport modes are fully integrated—note that private providers can still operate individual services, but to the client they all seem integrated.

MaaS does not need to apply to people only; it can apply to goods as well, such as for grocery delivery. Several companies benefited from the stay-at-home orders that were issued during the COVID-19 pandemic,

---

11. Collected from https://english.shanghai.gov.cn/en-Departments/20231215/60c9d3bfb6bc48 c48d2125cb2f0ea50f.html (accessed May 29, 2024).

including many MaaS companies that offered grocery and good delivery services. And I am not even going to speculate about the potential impact that autonomous vehicles could have on urban transport within the spectrum of mobility services.[12]

Beyond technology, the idea of MaaS originated from the concept of accessibility. For most of humanity, the dominant travel mode was walking. Cities were designed around pedestrians. Everything changed in the late nineteenth and early twentieth centuries, first with trains and streetcars, and then with private vehicles. All of a sudden, people could easily travel faster and farther, and the *pedestrian* was replaced by the *mobile* at the center of urban transport. The goal shifted to letting people go anywhere at any time, emphasizing the feeling of freedom that had not been possible before—hence the focus on flow. By the late twentieth century, the urban transport community changed its focus again. It realized that the feeling of freedom was just an illusion, as the freedom to travel anywhere often translated to being a slave to one's car and to having to drive long distances to access basic services. The goal shifted from being able to go anywhere to being able to access as many opportunities as possible in a reasonable amount of time. The *mobile* was replaced with the *accessor*. For example, how many stores are accessible within a fifteen-minute walk? How many jobs are accessible within a thirty-minute transit ride? With the *accessor*, people still have the freedom to go anywhere, but often they do not need to go excessively far, which both saves time and relieves overall traffic. Ideally, people also have several travel mode options.

Accessibility is related to Tobler's first law of geography: "everything is related to everything else, but near things are more related than distant things." In the late 1950s, Walter Hansen came up with a measure of accessibility, in which opportunities located farther away do not contribute as much to accessibility as opportunities located closer.[13] As a result,

---

12. At the time of this writing, there is just too much uncertainty.

13. The equation is as follows: The accessibility A_{im} of zone i using mode m is defined as $A_{im} = \sum_j E_j \cdot f(c_{ijm})$, where E_j is the number of opportunities in zone j and $f(c_{ijm})$ is a cost function to travel from zone i to zone j with mode m. For example, a typical cost function is $f(c_{ijm}) = c_{ijm}^{-b}$, where c_{ijm} is the travel time to travel from zone i to zone j with mode m, and b is a constant that can defer by

locations across a city can be compared based on their accessibility. For example, a location can enjoy high accessibility because grocery stores are accessible quickly, regardless of the travel mode selected (i.e., a quick car or transit ride or a short walk).

Reflecting on what we've learned about transport, accessibility is about being able to access many opportunities, regardless of the travel mode selected. And this is where MaaS is particularly relevant. By focusing on accessibility and mobility as opposed to traffic flow or single modes, MaaS can ensure that every resident has multiple options to reach a destination. But this is not possible without integrated transport.

From private, public, and active transport to integrated transport, transport infrastructure has evolved significantly over time and will continue to evolve in the future. We've learned about water and transport systems that have been integral to cities since the very first human settlements. Now let us examine a much more recent—and perhaps even more important—type of infrastructure, one that only came to be in the late nineteenth century. We travel from Shanghai to Cape Town to learn about electricity generation.

LYON'S BOUCHONS

"MERCI MARIE" reads a large sign on the hill next to the Basilica of Notre-Dame de Fourvière in Lyon (France) every year in December. My wife's first name is Marie-Agathe, but everyone calls her Marie. She is from Lyon. We often spend the winter holidays in Lyon, and the usual joke is to look at the sign, then at her, and thank her. I am sure all Maries in Lyon go through the same thing. It is the same joke every year, and it never gets old. The sign's origin begins in 1643, when a plague epidemic struck Lyon. As was usual at the time, the people prayed to a divinity—in this case, the Virgin Mary—and promised to pay tribute to her every year if the town was spared; it was. Since then, on December 8, the people of Lyon light candles and place them on windowsills. Since 1999, the city has scaled up the event

---

places and modes. Note the negative sign "–" in front of the constant b so that higher travel times result in lower values, and thus opportunities located farther away do not contribute as much to accessibility.

and has been organizing the Fêtes des Lumières (Festival of Lights) across the city over four days in early December. The festival has become a major attraction in Europe, one attended by several million visitors every year. Navigating the flow of visitors through all the shows, exhibits, and artwork has become a challenge that would not be possible without integrated transport.

As the third largest city in France, Lyon is located at the confluence of the Saône River and the mighty Rhône River (one of Europe's largest rivers)—the entire Rhône valley is particularly known for its wine. Lyon has always been known as a transport hub. It is located between Paris and Marseille (France's second largest city), and it is also close to Geneva in Switzerland and Torino in Italy. Transport runs in the veins of Lyon. It is an inland maritime hub, a railway hub, an expressway hub, and an air travel hub. In the context of this chapter, Lyon has much to teach us. It is an example of what to do and what not to do.

Louis Pradel was the mayor of Lyon from 1957 to 1976. He worked hard to modernize the city. On the positive side, he pushed for the construction and upgrade of many infrastructure systems, including Lyon's sewer system, the Part-Dieu financial center, hospitals, and a major conference center by the Parc de la Tête d'Or. On the negative side, his most controversial decision was to push for the construction of an expressway to cross the city center. After visiting several American cities in the 1950s, Mayor Pradel fought to have the major expressway that links Paris to Marseille go through the city center. A tunnel was built through the Fourvière Hill, and an expressway crossed the Presqu'île (i.e., the peninsula) through the Perrache neighborhood. In the end, the expressway resulted in higher congestion, lower air quality, and the division of the Perrache neighborhood—similar to what happened in many other cities like Boston and Seoul. In French, *bouchon* as featured in the title of this story means "cork" (e.g., for wine bottles), but the same term is used for traffic jams. The traffic in Lyon can get bad. In 2016, the expressway was downgraded from a national to a metropolitan expressway. The downgrading now gives the city authority to do what it wants with it. Like many cities in the world, Lyon is trying to get rid of its expressway. As of this writing, there are plans in place to change it into an urban boulevard and remove the interchange.

Lyon also offers a great example of what to do. Its urban transport system is particularly well integrated. The metropolitan area has a regional rail system, like most cities in France, called TER for *Train Express Regional*. Lyon's public transport system is run by the TCL for *Transport en Commun Lyonnais* (Lyon Public Transports), which operates 4 metro lines (started by Mayor Pradel), 7 tramway lines, 2 funicular lines, 26 high-frequency "C" bus lines, and over 120 regular bus lines. For a metropolitan area of only 1.7 million people, this amount of public transport infrastructure is amazing.

Many of the metro stations have strategically been placed to serve a maximum number of people. Some of them offer a great example of TOD design. All seven tramway lines opened in the 2000s. One of them, the Rhône Express, links Lyon's major train station to the Saint-Exupéry airport twenty-three kilometers (fourteen miles) away. The two funicular lines are older,[14] with one bringing people from low-lying lands by the rivers to the top of the Fourvière Hill. When I lived in Lyon in 2005–2006, the high-frequency "C" lines did not exist. I remember coming back one year and being amazed by how much more complex the metro map had become, and that was only thanks to the "C" lines. Some trunks of the "C" lines work like BRT systems, and the wheels are covered to give the impression that they work like rail systems. Does that remind you of anything? Solutions like the TransMilenio do not apply only to cities in lower- and middle-income countries.

In addition, Lyon is home to one of the first bike-share systems, the Vélo'v (which is well integrated with the TCL), and to more than one thousand kilometers of bike paths. The city is also enjoyable for pedestrians. Sidewalks tend to be large. Large footpaths were built on the embankments of the two rivers, where friends and families can get together to relax, take a stroll, or have a drink. At least two busy shopping streets have been pedestrianized, including the famous Rue de la République that links the Place Bellecour to the Place des Terreaux.

For drivers, the city has many large boulevards. The traffic is bad but not horrible. Finding parking is never easy, but there are many parking garages. In my experience, virtually everyone in the city has access to multiple travel options. They can drive if they want, take transit, bike, or walk. Again, Lyon has much to teach us when it comes to integrated transport.

In Lyon, a *bouchon* is also a traditional restaurant where they usually serve hearty food, like my personal favorite: andouillette. Wine is often served in a fifty-centiliter *pot* (small glass bottle), and when it is empty, you need to tell the server that you are *sur les graviers* (on the gravels) so you can get another. The expression comes from before the Rhône River was dammed downstream. It would occasionally dry out, and the gravel at the bottom would show.

Are you on the gravels yet? If not, let us move on to electricity generation.

---

14. In fact, they are older than London's first metro line. The Lyon people like to claim that they built and operated the first metro line, not London, although I am not sure a funicular line counts as a metro line.

PART THREE
ENERGY

CHAPTER TEN

Electricity Generation

The Powers That Power Cape Town

$\lightning$

CAPE TOWN.[1] WHEN I THINK ABOUT CAPE TOWN, THE TYPICAL POSTCARD picture of the city appears in my mind: Cape Town stadium, lush green fields at Green Point, blue-green water, the waterfront, Signal Hill, Table Mountain at the back, and, of course, perfect temperate weather. The Cape of Good Hope, despite its name, has offered rough seas to generations of sailors. One also cannot disassociate South Africa from Nelson Mandela, who, like Mohandas Gandhi,[2] fought for equality and who must be one of the most inspiring humans of all time.[3] Apartheid affected millions of Black people over the years—a stain that will forever stick to South Africa, reminding us how cruel, stupid, and counterproductive racism and discrimination are. Regarding infrastructure, South Africa suffered a devastating water crisis in 2018 when the levels of the main drinking water reservoir used by the city neared zero. The event was highly publicized around the world. Not as publicized, but nonetheless

---

1. What is the capital of South Africa? I figure most people would say Cape Town (including me). The geography buffs would be quick to correct them and proudly announce that the capital of South Africa is Pretoria . . . and they would be wrong, too. It turns out that South Africa has three capitals: Pretoria is the administrative capital; it is where the president resides. Cape Town is the legislative capital; it is where the Parliament is located. And Bloemfontein is the judicial capital; it is where the Supreme Court is found. South Africa is the only country on Earth with three capitals.
2. Both men were imprisoned in the same prison, Constitution Hill, more than fifty years apart.
3. After the COVID-19 pandemic, while I lived in Paris in summer 2020, posters with health workers were featured around the city with one of Nelson Mandela's most famous quotes: "I learned that courage was not the absence of fear, but the triumph over it."

deserving attention, is the electricity crisis that has affected not only Capetonians but all South Africans.

Have you ever heard the term *load-shedding*? It is a term that South Africans know too well. Load-shedding occurs when the electric utility company shuts off electricity supply on purpose, most often according to a predetermined schedule, because the demand for electricity exceeds the supply. Usually, areas of a city are affected one at a time, on a rotation. In 2007, South Africa entered an electricity crisis. Fueled by decades of Hollywood-style corruption[4] that continued well into the 2010s, the electricity infrastructure in South Africa, run by the company Eskom, which supplies about 95 percent of the electricity, had been severely neglected and could not keep up with an increasing demand for electricity.

From being almost nonexistent by the mid-nineteenth century, electricity has arguably become the most important type of infrastructure. Virtually every activity we engage in requires electricity, and our society has become as dependent on electricity as our bodies are dependent on water. It is almost impossible to imagine a world without electricity. (I write "almost" because a Japanese film director did just that and directed the 2016 movie *Survival Family*, which offers a disconcerting tragicomedy of a world without electricity.)

From being generated to being available at the flip of a switch, electricity is a fascinating realm. First, we must remember that, except for lightning and a few other natural events, electricity does not exist as a raw form of energy in the world. Unlike water, electricity cannot be collected. Instead, it must be generated, and its generation requires that one form of energy available in nature be converted into electricity. This conversion entails important efficiency losses, but there are many sources of energy in the natural world, which means there are many ways to generate electricity.

What is the best way to convert one type of energy into electricity? In 1831, the British scientist Michael Faraday found that if you rotate a

---

4. Involving McKinsey (an internationally renowned U.S. management consulting firm) and the Guptas (an obscure family with ties to the president of South Africa) in a scenario that inevitably led to corruption.

copper disk between the two legs of a horseshoe magnet, you can create an electric current. The keyword here is *rotate*. Nowadays, the most typical way to generate electricity is to have electromagnets rotate on a shaft (i.e., a rotor) with coils of wire placed around it (i.e., on a stator). One name for these types of generators is *turbine*.[5] As the electromagnets rotate, a current is generated in the coils in the stator. This is electricity. The concept is relatively simple, and, importantly for us, we see that the raw energy that we need is used to *rotate* the rotor. To be more precise, we need *mechanical* energy to provide movement, which will be converted into *electrical* energy. The question therefore becomes: Where do you get this mechanical energy?

When we look in the environment, the best places to find mechanical energy are in the wind and in rivers, and we use both to generate electricity. With wind turbines, large blades are attached to an electricity generator to generate electricity. When many turbines are present, we have a wind farm that can produce significant amounts of electricity. The concept is fairly simple, and wind turbines can be installed wherever there is wind, whether onshore or offshore.

With rivers, there are at least three general options to generate hydroelectricity.[6] The most common is to build a dam with turbines at the bottom that mainly use the hydraulic pressure from the drop in elevation to generate electricity. This option includes the flooding of large areas upstream to create a reservoir. A second option is to place a turbine directly in the river that mainly uses the force of the flow to turn the turbine (sometimes smaller dams are built as well). These are called run-of-the-river (ROR) power plants. They are less popular in part because they require the right conditions (i.e., a strong enough and consistent flow) and because how much electricity they can generate depends on the seasons (more in the spring with heavy rains and snow melt, and less in the summer and winter). On the plus side, they do not flood large areas

---

5. We have all seen the terms *rotor*, *stator*, and *turbine* at school, and I know many readers will not necessarily be fond of them. Don't worry, the technical content will be over soon.

6. *Hydro* for water.

upstream, but it also means that they cannot help as much to regulate the river depth downstream (sometimes important for navigation) and they are not as effective for flood protection. We will learn about the third option later. It has a twist. We must be patient.

Because the energy available from wind and rivers is already in the form of mechanical energy, these processes are highly efficient (typically above 90 percent efficiency). Other forms of natural mechanical energy include the tide and waves, but, at the time of this writing, they are not as widely used.

Another option is to harvest a different type of energy and initially convert it to mechanical energy, followed by electrical energy. One popular form of energy that is used in practice is chemical energy. In essence, chemical energy in some medium (e.g., wood) is used to produce thermal energy (i.e., heat) that is itself used to produce mechanical energy. Some of the most concentrated forms of chemical energy available are stored in fossil fuels that burn easily. There are a couple of ways to generate electricity from burning fossil fuels. Like fossil fuel–powered vehicles, some small power plants use fossil fuels like diesel to actuate a piston to turn a rotor. In contrast, coal- and some oil-fired power plants, as well as nuclear power plants, use the heat generated to produce steam at high pressures that is then used to operate a turbine. We can also burn natural gas directly to rotate blades like jet engines in an aircraft that in turn rotate the shaft in the turbine. All systems that use heat are called *thermal systems*. They also include biomass power plants that burn crops and solid waste to generate electricity as well as geothermal and solar thermal plants. Geothermal power plants are best constructed in areas that possess geothermal activity (i.e., underground volcanoes), as is the case in Iceland. In contrast to solar photovoltaic panels (discussed below), solar thermal plants use solar energy to create steam to operate a turbine here as well.

The use of heat generally works well, but efficiencies now drop significantly since the energy is converted three times—first from chemical to thermal, then from thermal to mechanical, and then from mechanical to electrical. Efficiencies range from about 10 percent for geothermal (because the ground is not that hot) and about 20 percent for solar thermal to between 30 and 45 percent for coal, oil, natural gas, and nuclear power plants. That said, because most of the wasted energy is in the form of heat, some power plants recuperate that heat and use it. Some natural gas power plants use it to create steam to operate a second turbine; these plants are called *natural gas combined cycle* (NGCC) power plants. Some thermal power plants also use the waste heat to heat buildings in the winter or even to cool down buildings in the summer.[7] These power plants are called combined heat and power (CHP) plants, and the heat can be distributed to a network of nearby buildings in what are called *district heating and cooling systems*. This is typically the case for large building complexes. In Saint Pierre, my hometown, electricity is generated by six diesel motors, and the waste heat is used to heat administrative buildings around the island. The system works well. If we account for the energy recovered by using the heat, efficiencies can climb up to 80 percent.

In addition to chemical energy, other forms of energy can be harvested. The most famous alternative is solar photovoltaic energy, which uses solar energy directly to generate electricity. It is not the thermal energy contained in the heat that solar panels use but the solar energy contained in the photons—usually more photons translate to higher temperatures, so warm places with a lot of sun tend to be good candidates for solar photovoltaic energy. At the time of this writing, efficiencies for solar panels remain relatively low (around 20 percent), but the sun is present virtually everywhere. Moreover, solar panels are scalable in the sense that it is easy to install a few panels on one's roof or build large solar farms.

The last form of electricity generation we should discuss is different because it is not a real method to generate electricity; however, it has become key to transforming the power grid. It goes by the name "VPP," which stands for virtual power plant. VPPs are not fake power plants located on the internet; they are electricity storage facilities that can act as power plants during peak demand. As we will see, at every instant, the

---

7. Although it sounds counterintuitive, adsorption chillers use heat to chill water.

amount of electricity supplied must match the amount of electricity consumed. For the longest time, this constraint was one of the biggest drivers for massive power plants that could be controlled easily, but we are getting better at storage. Plus, more and more people want to have better control over their electricity generation capabilities—for example, by installing solar panels on their roofs. With electricity storage, even if the electricity is not needed right now, it can still be produced and stored for later. Storing electricity is tricky, however. There are no natural reserves of raw electricity that can be leveraged.[8] The best solution is to convert electrical energy back to another form of energy, store it, and then convert it again into electrical energy when needed. It is not ideal because of efficiency losses, but it works, and a series of solutions exist.

The most popular electricity storage method is the rechargeable battery that stores chemical energy. When we produce an excess amount of electricity, it can be converted to chemical energy that can be converted back to electricity when needed. Batteries work well, but they tend to be on the small side. Manufacturing batteries is also not terribly environmentally friendly. While batteries have a bright future, especially to be used in cities, other methods also exist. One of them is to pump water up. Here, we need two reservoirs at two different elevations, one higher and one lower.[9] When water flows from the higher reservoir to the lower one, electricity is produced in a turbine as is conventionally done in hydroelectric dams, but when there is an excess of electricity, water can be pumped from the lower reservoir to the higher one. Here, the excess electricity generated is consumed by the pump, and electricity is then generated later when the water flows down again. This is our third option to generate hydroelectricity—I said there was a twist.

Another electricity storage method is to compress air or liquid in a container or underground using electricity and later use this compressed air or liquid to turn a turbine when needed—it works a little like the

---

8. In contrast, storing treated water is easy. This is why water can be treated at a continuous rate and stored in a reservoir when not needed.

9. In dry places like Saudi Arabia, they are looking for solutions that use sand.

artesian wells that we learned about in Rome, with flows so strong that electricity can be generated. Another simple approach is to use electricity to rotate a large fly wheel that continues to rotate because of inertia even when we stop powering it, and it can produce electricity.

We will talk more about electricity storage, but the point here is that we are getting better and better at storing electricity. In fact, we are getting so good at it that when all these storage devices feed electricity into the grid, they act as a VPP.

The next step is to compare the different electricity generation methods that we have listed and try to determine which methods are preferred. We can compare them based on at least three criteria: efficiency, operating characteristic, and sustainability.

First comes efficiency. Any conversion from one form of energy to another generates waste; that is the second law of thermodynamics.[10] Let us break things down a little. Note that I did not write that energy was *lost* in the process. As the eighteenth-century French chemist Antoine-Laurent de Lavoisier said, "Nothing is lost, nothing is created, everything is transformed"[11]—this is the law of conservation of energy.[12] But when you convert mechanical energy to electrical energy, some energy will inevitably be wasted as heat (i.e., converted as thermal energy); that is what efficiency is about. When little energy is wasted, efficiency is high. Conversely, when much energy is wasted, efficiency is low. Naturally, methods that are more efficient are preferred. It should not be a surprise to learn that methods that involve fewer energy conversions tend to be more efficient. This is why wind, rivers, tides, and waves that are already in

---

10. A system will always go from a state of higher equilibrium to a state of lower equilibrium unless you put work into it. In other words, it will always go from more orderly to less orderly unless work is added to control the order. Here, the energy is first in a state of high equilibrium, and converting it to another form of energy inevitably destabilizes this equilibrium. This is why some energy is wasted along the way, and it is wasted as heat. This is also why the field is called thermodynamics—*thermo* for heat.

11. Or, in its original French, "Rien ne se perd, rien ne se crée, tout se transforme."

12. In thermodynamics, this is also the first law—that is, conservation of energy—versus the second law defined above.

the form of mechanical energy are preferred since they are 90 percent efficient.[13]

When chemical energy is used to generate electricity, natural gas is preferred since natural gas power plants tend to have efficiencies around 40 percent in contrast to 30 percent for oil and coal. Nuclear power plants also have an efficiency of around 30 percent. Biomass power plants have lower efficiencies, around 25 percent, because the chemical energy in the fuel (e.g., crops or solid waste) is not as dense as fossil fuels.[14] Finally, we have solar thermal and geothermal power plants as well as solar photovoltaic farms. Note that we can give thermal systems an efficiency boost as well by recuperating and using the wasted heat (i.e., thermal energy) for other applications, such as heating buildings (as is the case with CHPs) or producing more electricity (as is the case with NGCCs). The only exception to this efficiency argument is when you want to create heat in the first place—by using an electric radiator, for example,[15] in which case, you want as much waste heat as possible.

Second comes operating characteristic. By that, I mean whether we can produce as much electricity as we want or whether we are limited by some factor like the weather. The technical terms here are *flexible* and *intermittent*. Let us go back one step. As mentioned before, at every instant, the power grid must be able to supply exactly what is being demanded. If more electricity is supplied than demanded, the voltage increases, which can damage electrical equipment and lead to a blackout. If less electricity is supplied than demanded, the voltage decreases, and electrical equipment does not work as designed, which can also lead to failure (i.e., blackout) or something less

---

13. And where does the remaining 10 percent go? It is wasted as heat of course.

14. That is why fossil fuels are so great. They pack a high amount of energy per unit volume that can be used to generate electricity. But they exist in finite quantities on Earth, and they generate abundant greenhouse gases. Burning fossil fuels in small quantities is fine, but having an entire society based on the burning of fossil fuels inevitably leads to dramatic consequences. We will discuss sustainability later.

15. Using gas to heat buildings is not as efficient as using electric radiators because some of the heat goes up the chimney and is not used. Electric radiators are therefore technically preferable, but—and there is a *but*—it depends on how electricity is generated in the first place. If electricity is produced with natural gas that has an efficiency of about 40 percent, it is preferable to have a gas furnace with an efficiency of 85–95 percent to heat a building.

severe (like a brownout; think about a light bulb getting dimmer). Because the demand for electricity changes during the day, matching consumption exactly can be challenging. The real limiting factor is the maximum peak power than can be supplied. That peak power usually happens on a hot summer day in late afternoon, when air conditioners are on at the same time as TVs, lights, water heaters, computers, and all the other appliances that we usually run when we arrive at home. Then, at night, the demand will be much lower. To accommodate such a dramatic increase and decrease in electricity demand, *flexible* power generating methods are preferred, such as hydroelectric power plants, nuclear power plants, biomass power plants, fossil fuel–powered power plants, and so on.

With intermittent methods like solar and wind farms, we cannot control how much power is generated simply because we do not control the weather. Flexible methods are therefore preferred to intermittent methods, at least for now, but conditions are changing. As mentioned, we are getting better at storing electricity, and all these storage devices combined can act as a VPP. We are also getting better on the demand side. While we cannot control when people switch their appliances on or off,[16] demand response strategies encourage and pay people to use less energy during peak times, like on a hot summer day in late afternoon (e.g., by encouraging customers to increase their thermostat by a few degrees). In some cities, electrical utilities even have direct control over these appliances. For example, a utility may increase the thermostat of its customers directly and pay them for it,[17] just to make sure they can handle peak demand smoothly and do not have to rely on firing up an old coal-powered plant.

Finally, the third criterium is sustainability. What does sustainability mean? The term is so popular and trendy that most people often confuse it with other concepts. But what does *sustainability* really mean? Namely,

---

16. As a funny anecdote, the people of the United Kingdom love tea. I have read that during some prime-time shows, the number of households that turn their electric kettles on during a commercial can be so high that a surge in electricity demand is noticeable.

17. Yes, customers get paid for the electricity they do not consume. The unit used sometimes is called a *negawatt* for negative watt consumed.

I like to take the literal meaning of sustainability, as in, if something cannot be sustained in the long term, it should not be done. For example, spending more money than what you earn is not sustainable. Back to electricity: because it is not available in raw form, we need to harvest other types of energy, and therefore we should see whether the harvest of these other types of energy is sustainable. Rivers, the wind, the sun, waves, and tides are here to stay, so we can harvest them as much as we want—this is why they are also referred to as *renewable energy*. Geothermal energy is also a type of renewable energy. What is not renewable? Fossil fuels, since it took hundreds of millions of years to produce them. They are by definition not sustainable, and once we burn them, they are gone. The same principle applies to nuclear energy that requires uranium or plutonium as its fuel, which, once processed, is gone forever. Using fossil fuels and nuclear energy is therefore like spending more money than we earn. Electricity generation methods that are sustainable are preferred.

We need to push this argument further because the burning of fossil fuels has a double impact. Not only is using fossil fuels unsustainable in the first place, but burning them also emits greenhouse gases that end up in the atmosphere. And we have burned such a high quantity of fossil fuels and emitted such a high quantity of greenhouse gases that we have altered the climate—that is why, despite being two very different concepts, sustainability and climate change have become directly linked. While the Earth can naturally absorb some of the greenhouse gases emitted, we are simply emitting too many. This is why we could burn coal for centuries without changing the climate, but now the rate at which we are burning it is unsustainable from a greenhouse gas point of view as well. Burning fossil fuels is therefore unsustainable not on one count but on two—a similar observation can be made with cars.[18] And—here is the cherry on top—the greenhouse gases that are generated because we burn

---

18. If you recall the chapter on traffic and roads in Los Angeles, the story with cars is not that different. Cars that run on fossil fuels are unsustainable not only because they require fossil fuels but also because they require more space than what is available in streets. Fossil fuel–powered vehicles are therefore unsustainable on two counts as well.

fossil fuels also severely impact air quality and present a public health hazard to humans. These are pretty strong arguments against burning fossil fuels. This is also why nuclear energy is preferable to fossil fuels.[19] Moreover, nuclear power plants are known to be ultra-safe.[20] This is not to say that nuclear power plants are perfect. Not only are they not sustainable by nature, but the byproducts of nuclear power plants (i.e., radioactive waste) are also difficult to manage and have unknown long-term impacts. But in the medium term, nuclear is preferable to fossil fuels.

In the end, which electricity generation method is best? The golden child is hydroelectricity. It is not perfect, but it is as good as it gets since the raw energy form is already mechanical energy (i.e., it is efficient), it is a flexible means of generating electricity, and it is sustainable. This is why the seven largest power plants in the world are hydroelectric power plants,[21] with the Three Gorges Dam coming in first place (see story at the end of this chapter). Nevertheless, we have nearly maxed out and built hydroelectric dams almost everywhere we could (at least in high-income countries). Next comes all renewable energy methods, from wind, geothermal, solar thermal, and solar photovoltaic. Knowing which method is best depends on the local context. Places with a lot of wind should build wind farms, places with a lot of sun should build solar thermal power plants or solar photovoltaic farms, and places with geothermal activity can build geothermal power plants. Moreover, we can supplement intermittent electricity-generation methods with the least unsustainable methods such as nuclear and biomass.[22] Plus, with

---

19. Which is why Germany's decision in the 2010s to move from nuclear to coal to generate electricity is absurd and not environmentally friendly. But the country is also phasing out coal at the time of this writing.

20. In fact, they are so safe that the level of radiation in nuclear power plants is lower than in coal power plants—every object emits some radiation, and nuclear power plants are so safe that they emit less radiation that most places.

21. According to https://en.wikipedia.org/wiki/List_of_largest_power_stations (accessed May 29, 2024).

22. We have not discussed biomass much. Is it sustainable or unsustainable? Technically, it can fall under either category. Burning crops that grow anyway and that capture CO_2 from the atmosphere is relatively sustainable. Burning plastics or other solid waste is not particularly sustainable because we should not generate much solid waste; we will learn more about solid waste when we travel to Tokyo.

advances in electricity storage, VPPs should take a more prominent place in the future. In fact, many expect the massive transition to electric vehicles to offer enormous opportunities (all these vehicles will make up several VPPs). Finally, if we must go the fossil fuels route for the time being, natural gas is preferred since it emits fewer greenhouse gases. In terms of greenhouse gases, the worst method is coal.

Now that we have learned about the forms of electricity generation, let us get back to Cape Town: How is electricity generated there? To my despair, the main electricity generation method there is coal. South Africa has large reserves of coal, and it has decided to fully utilize that supply to generate electricity.[23] In 2018, the South African Department of Energy reported that 72 percent of the electricity was generated from coal. This number is followed by 7 percent from gas, 5 percent from pumped storage (i.e., pumping water to a higher reservoir when an excess of electricity is produced), 4 percent from hydroelectric dams, 4 percent from wind, another 4 percent from solar (1 percent from solar thermal and 3 percent from solar photovoltaic), 3 percent from nuclear, and finally 1 percent from a mix of methods, including biomass. This electricity generation mix leaves a lot to be desired, and serious investments seem to be made in wind and solar. I have no doubt South Africa can do much better.

South Africa has another specificity. At the beginning of this chapter, we discussed the practice of load-shedding. It is essentially a strategy to control the demand. As we have seen, we can control how much electricity is produced, but we cannot produce more electricity than the amount all the power plants can produce, and if the demand exceeds the supply, the electricity grid can get severely damaged. In an effort to control the demand, Eskom practices load-shedding so that the amount of electricity that is demanded never exceeds what the power plants can produce.

---

23. And, as I have repeatedly written in this book, diversity is always good. Basing a large part of an economy and most of the electricity sector on one energy source is problematic, and it will most often lead to negative consequences unless more diverse electricity generation methods are added to the mix. The 2022 Russia-Ukraine war is a good reminder of this phenomenon since some European countries were too heavily tied to Russia as a large energy provider.

Once electricity has been generated, the next step is to distribute it from power plants to wherever that electricity is consumed through an electrical grid. We have alluded to the grid a few times, but there is much more we need to learn. For this, we will cross the Atlantic Ocean and travel to Chicago.

TAMING THE MIGHTY YANGTZE RIVER

The Yangtze River is mighty. It receives water from one-fifth of China. From its source in the Tibetan Plateau, it travels over more than six thousand kilometers (close to four thousand miles) until it meets the East China Sea in Shanghai. Many large cities were established and historically lived off the Yangtze River, including Chongqing, Wuhan, and Nanjing. Downstream of Chongqing, the mighty Yangtze is so strong that it carved down the earth, forming a series of three majestic gorges: the Qutang Gorge, the Wu Gorge, and the Xiling Gorge.

To an engineer, the Yangtze River is a formidable source of energy that can be harvested, a beast that must be tamed. Plans to install a hydroelectric power plant on the Yangtze River have been entertained throughout the twentieth century. Mao Zedong even wrote a poem about it. We had to wait until 1994 for construction to begin. It ended in 2012. The result is simply the largest power plant in the world. Its name: the Three Gorges Dam.

While building a dam in a large river is never a small affair, we are now talking about the mighty Yangtze River. The dam that was built is 2.34 kilometers long (1.45 miles) and has a maximum height of 185 meters (607 feet). It is gigantic. It has close to half a million metric tons of steel and thirty million cubic meters of concrete[24]—that is close to nine times more than the Hoover Dam located at the border of the U.S. states of Nevada and Arizona. The dam is said to have required a labor force of forty thousand workers for its construction.

The dam has thirty-two 700-megawatt turbines and two smaller 50-megawatt turbines, for a total capacity of 22,500 megawatts. It is huge! The Three Gorges Dam is not the world's largest dam—that title goes to the Tarbela Dam in Pakistan— but it is the largest power plant in the world. The second and third largest power plants in terms of power capacity are also hydroelectric power plants, but they do not compare to the Three Gorges Dam. The second largest is the Baihetan Dam, also in China, and has a capacity of 16,000 megawatts. The third largest is the Itaipu Dam, located at the border of Brazil and Uruguay, and it has a capacity of

---

24. From the Wikipedia page "Three Gorges Dam," available at https://en.wikipedia.org/wiki /Three_Gorges_Dam (accessed May 29, 2024).

14,000 megawatts. Currently, the largest thermal power plant in the world is a natural gas power plant located in the United Arab Emirates. With a power generation capacity of about 8,700 megawatts, it is the eighth largest power plant in the world.[25]

Such a large project as the Three Gorges Dam is not without faults. Its construction meant the flooding of 632 square kilometers (244 square miles) of land over a length of 600 kilometers (373 miles) upstream. This flooding is estimated to have required the displacement of more than 1.3 million people living in about 1,500 cities. Moreover, many treasured historical and archeological sites were submerged by the rising waters. From an environmental perspective, dams always present several problems. Besides the impacts on the local wildlife, sediments[26] that are normally flushed out downstream of the river are now blocked by the dam. To partly mitigate this problem, the dam includes massive sluice gates.

On the positive side, the electricity generated comes from a renewable source of energy (electricity that may very well have been otherwise generated by burning coal). The dam also helps control the flow of the Yangtze River, therefore serving as flood protection for cities downstream. Moreover, damming the river made navigation from Shanghai to Chongqing possible (a more environmentally friendly alternative than using trucks or trains for freight transport). On the northern end, the dam features a five-lock system (the largest in the world) to enable ships to go up and down the river. For smaller boats that want to cross the dam more quickly, a boat elevator system was built.

As discussed in the chapter, hydroelectricity is the golden child of electricity generation. The Three Gorges Dam is not perfect, but humanity's thirst for power has never been greater, and it will likely increase in the future. Tapping the renewable and freely available mechanical energy that nature offers is the best we can do . . . at least for now.

---

25. From the Wikipedia page "List of Largest Power Stations" available at https://en.wikipedia.org/wiki/List_of_largest_power_stations (accessed May 29, 2024).

26. Including those generated through the erosion of the bed of the Yangtze River and its tributaries due to the force exerted by the flow of water, which further contributes to the digging of the gorge.

CHAPTER ELEVEN

Electricity Distribution

The Birth of the Grid in Chicago

CHICAGO! THE WILLIS TOWER,[1] THE BEAN,[2] THE CHICAGO RIVER, LAKE Michigan, the Art Institute of Chicago. Home of the Bears, the Cubs, the White Sox, the Blackhawks, and the Bulls. The city that was largely destroyed during a massive fire in 1871, and the city that dyes its river bright green for St. Patrick's Day. In addition, Chicago hosted the World's Columbian Exposition in 1893, which is often seen as the birthplace of modern planning (emphasizing hygiene and showcasing the power of electricity). Chicago's solution to dealing with its drinking water and wastewater is also a saga with never-ending twists that includes raising most buildings in the city nearly two meters, reversing the flow of the Chicago River, and building gigantic underground tunnels to store wastewater.

Throughout the world, Chicago is perhaps best known for its gangsters and its speakeasies from the time of Prohibition, often personified by a single character: Al Capone. Its historical reputation as a corrupt city is not undeserved. In fact, the nickname the "Windy City" refers not to the weather but to the fact that politicians would sway their opinions easily.

---

1. I know, people in Chicago still call it the Sears Tower, but the name changed in 2009. For a country that nurtures a culture of innovation, Americans do not seem to like change that much. The imperial British system is still used in the United States, while the rest of the world uses the metric system. Enough with inches and feet; adopt the meter already!
2. Its official name is Cloud Gate, but it really does look like a bean.

Even today, Chicago carries an identity of violence and uneasiness,[3] which is again not undeserved. Like most U.S. cities, and symptomatic of deep unresolved issues in American society, the city is heavily segregated. The social tension is palpable. Although I had visited the city before, I had no idea what to expect when I moved here in 2012. The move to the United States, and to Chicago in particular, was not as easy as I had expected, but I came to love my life here. In the end, Chicago is a wonderfully complex city that has much to offer. Personally, I live on the north side, and one of my favorite activities in the city is to walk along the lake from my home to the city center.

Importantly for us, Chicago is also the home of the first electrical grid ever built. The *electrical grid* or *power grid* (or just *grid* for short) is the name given to the infrastructure that distributes electricity from power plants to wherever it is consumed. And perhaps to remain true to the nature of Chicago, the original grid was built by a dubious figure: Samuel Insull.

In the previous chapter, we learned how electricity is generated. Now let us see how it is distributed. There is one thing that we did not discuss in the previous chapter, and that is where power plants are located. After all, this book focuses a lot on urban infrastructure, but power plants are rarely located in cities themselves. Instead, they tend to be located next to the natural resources that they exploit. Hydroelectric dams are located on rivers. Wind and solar farms are located where it is windy and sunny. Coal plants can be located nearby coal mines—at least in South Africa. More accurately, thermal power plants that produce steam in general (e.g., coal and nuclear) tend to be located close to natural bodies of water (e.g., rivers, lakes, and even oceans), not to produce electricity but because they require a lot of water for cooling.[4] As a result, infrastructure needs to be

---

3. In your favorite search engine, start typing "why is Chicago so" and see the automatic suggestions that pop up.

4. The water used for the steam is in a closed-loop system and is reused. Pure water is used not to corrode the pipes and turbine blades. The water collected from water bodies is used in heat exchangers.

built to transport electricity, sometimes over hundreds of kilometers, and this is where the grid takes form.

Here, we can view electricity like water. A lot of electricity is produced in one place, and it needs to be distributed to many locations, but, instead of pipes, we have wires. These wires are typically divided into two categories: *transmission* lines and *distribution* lines. The transmission lines are the very large ones that we can see hanging from large metallic structures, often next to highways. Distribution lines are the smaller wires that we can see in city streets and back alleys on individual poles.[5] Transmission lines typically carry electricity with voltages over 200,000 volts versus less than 30,000 volts for distribution lines; power lines that carry voltages in between are called *subtransmission* lines. These high voltages contrast to the voltages provided by a typical outlet—between 100 and 240 volts depending on where you live. It is around 120 volts in North and Central America and around 230 volts almost everywhere else; a notable exception is Japan, where the voltage is only 100 volts.[6] Going from 200,000 volts to 100 volts is quite a transformation. But why do we even want to change the voltage across power lines in the first place?

I promise to make this quick. Voltage is a difference in electric potentials between two points (like pressure in water). Current is a flow of electric charge (like flow rate in water). Resistance is akin to a *consumption* of electrical energy; sometimes we want the consumption to be high (e.g., when we want to produce heat, converting electrical energy to thermal energy), and sometimes we want it to be low (e.g., in power lines where resistance is akin to friction losses in water pipes). Finally, power is the instantaneous amount of electricity that is produced and sent to the grid; it is a rate, not an amount. Power is energy per second; the unit is the watt. Energy is the total amount of electricity produced, sent, and consumed—for example, in

---

5. Although we see them less and less since they tend to be buried in the streets now.

6. Despite being a little more hazardous, a higher voltage is generally preferred because less material is needed in the wires (since the current, i.e., the flow rate, is lower) and the wires are therefore cheaper. This is why Europe went with 230 volts historically.

a month or in a year—and the unit is the watt-hour and not "watt per hour," as I too often see.[7]

When distributing power, the goal is to minimize the losses so that as little power as possible is wasted in the transmission and distribution lines. Because the resistance of a power line is a property of the material, we do not have much control over it. Instead, we can minimize the losses by having as low a current as possible, which translates into a high voltage.[8]

To transform the voltage, substations are located along the grid in the form of large transformers that transform electricity with a high or low voltage into whatever is desired. The technical terms are to *step up* or *step down* a voltage. Even at the turbine stage, when electricity is first generated, the voltage is relatively low (often around 22,000 volts), and it needs to be stepped up before it is sent over the grid. Transformers are large, often gray colored, looking a little like electric radiators with their metallic grates.[9]

The fact that we need to step up/down the voltage is also the reason why most appliances use alternating current (AC) power as opposed to direct current (DC) power. There was a famous and ferocious battle in the electricity world between Thomas Edison and Nikola Tesla that is well documented and was made into several movies.[10] Edison created the light bulb and built the first ever power plant—the Pearl Street Station in New York City in 1882. He supported the adoption of DC power. In contrast, Tesla, the electricity genius, supported the adoption of AC power. For a

---

7. Making sure people say "watt-hour" and not "watt per hour" has been a personal battle for me. Energy is typically quantified in joules. Power is energy per second, and it is therefore in joules per second, which is the watt. Somehow, instead of going back to joules for energy, some people decided that it is easier to multiple power by the number of hours it was consumed, so watt times hour (or watt-hour) as opposed to watt per hour, but one watt-hour is just thirty-six hundred joules. Tell your friends and family! Or demand to remove the unit watt altogether and solely use joules for energy and joules per second for power.

8. Power loss P_{loss} can be expressed by $P_{loss} = R \cdot I^2$, where R is resistance (which is a constant and a property of the wire) and I is current. Because we cannot control R besides selecting a material that has as little resistance as possible, we can try to make I as low as possible to make P_{loss} as low as possible. And making I as low as possible translates into a high voltage V since $V = P/I$, and P is fixed; it is the power output from the power plant.

9. Used for cooling since substations can get hot. Why? Because of conversion losses that are expressed in the form of heat, as we have seen.

10. Including the 2017 movie *The Current War*.

given voltage, DC power generates fewer losses than AC power, but it is difficult to transform the voltage of DC power, making long-distance electricity transmission impractical. In contrast, despite higher losses, it is extremely easy to transform the voltage of AC power. While Edison recommended keeping a constant voltage and building power plants every few kilometers, Tesla eventually won the battle, and we are all using AC power today.[11] The adoption of AC power and the transmission of electricity over long distances also significantly favored the adoption of electricity, even in remote places. In Chicago, at the time of the World's Columbian Exposition in 1893, electricity was available to few buildings contained in the city center. With the adoption of AC power, combined with the business acumen of Samuel Insull, nearly every building in the entire Chicagoland area had access to electricity by the 1920s.

Originally from England, Insull became Thomas Edison's secretary in the 1880s when he came to the United States. Rising through the ranks, by the 1890s he asked to become president of the then-small Chicago Edison Company.[12] Initially, individual electric utilities both generated and distributed electricity.[13] In Chicago, little by little, Insull bought out all his competitors. It was by merging the neighborhood grids together that he developed the first regional power grid. An astute businessman, Insull analyzed when his customers were consuming electricity. What he realized is that the demand for electricity was uneven, with peaks in the early evening. The *peak load* (i.e., the maximum demand at any time) is the limiting factor of an electric utility; this is why load-shedding is practiced (remember Cape Town). By analyzing how his customers consumed electricity, Insull could try to bump up the consumption of electricity in off-peak hours, therefore generating additional revenues without

---

11. A few high-voltage direct current (HVDC) power transmission lines exist in the world because they generate fewer losses. But building them makes financial sense only when distances between them are extremely long.

12. Insull even took a significant pay cut because the Chicago Edison Company was small at the time and had few customers. Today, the name of the company is Commonwealth Edison (ComEd), and it is still the main electricity distributor in the Chicago region.

13. This is not the case anymore in many countries. Different companies generate and distribute electricity to avoid monopolies so that even small power plants can sell their electricity.

requiring more equipment. One particularly clever strategy of Insull's was to first work with and then buy local train companies that ran on electricity. When people are traveling on the train, they are not home consuming electricity, and the two therefore complement each other. In essence, he was partly controlling the *load* (i.e., the demand for electricity). In addition, he could use the land by the rail tracks to install transmission lines to reach a larger market. In the end, by creating and buying out a significant number of companies, Insull built a business empire that required dubious transactions to stay afloat during the Great Depression and ultimately led to his fall.[14]

Back to the power grid. We now know that the grid is composed of power plants, transmission lines, distribution lines, and substations to step up/down the voltage. There is one more important piece of the puzzle. Look out the window, go outside, or think about the last time you walked/rode/drove by power lines—how many power lines next to one another were there?

Seriously, go look or think about it: How many lines were there?

My guess is that there were three or six power lines. If we recall how turbines work, we remember that we are rotating electromagnets on a rotor with coils of wires around them. With this configuration, we can easily generate more than one single current. The general practice in the entire world is to produce three currents that are out of phase with one another by 120 degrees;[15] phases are proper to AC power since the current is alternating, as the name suggests (from positive to negative, following a sinusoidal form). It is called three-phase power. This is why you see three power lines on electric poles and transmission towers, one for each phase. Some transmission towers have six lines—that is, two lines for each phase because too much power is transmitted to fit in one single line per phase. At the very top of transmission towers, an additional wire is

---

14. I recommend John Wasik's 2008 book *The Merchant of Power: Sam Insull, Thomas Edison, and the Creation of the Modern Metropolis* to learn more about Samuel Insull.

15. And the sum of the voltages is always zero, so a neutral line is not necessarily needed. This is why there are generally three lines and not four.

usually present as lightning protection. On individual poles in cities, phone and other telecommunication cables are sometimes present as well, generally below the three electric cables because they are safer (i.e., they do not carry high electric currents). If you see a pole with only one or two lines, it is probably not used to distribute electricity (it may be a telephone line).

More than three currents could easily be produced, but three has become the convention. We need to remember that in addition to turbines, the grid is composed of thousands of kilometers of power lines and thousands of substations. Each current requires an individual power line and an individual transformer to be stepped up/down. Producing more currents would incur a significant increase in cost.

If you are like me, whenever you see power lines, you will now count how many there are. I have seen them everywhere, from Chicago to small rural villages in Vietnam, and I systematically count three. Seeing these three juxtaposed power lines has a calming effect on me. It gives me a sense of stability and uniformity, no matter where I am. It is the one constant in this tumultuous world, which is a little ironic when you know that the electrons in the power lines are endlessly going back and forth fifty to sixty times per second (they are everything but constant). This going back and forth is called frequency. It is the number of times the current alternates per second, which is measured in hertz,[16] named after the German physicist Heinrich Rudolf Hertz. The frequency is sixty hertz in North America and fifty hertz almost everywhere else in the world.

When it comes to distribution, the three currents are dispersed evenly across a city so that more or less the same power is consumed from the three phases. In countries that operate at around 230 volts, buildings are generally supplied with one phase only; electrical panels are laid out horizontally on one or more levels. In countries that operate at around 120 volts, however, two phases are generally supplied. This is why electrical panels in North America are divided into two vertical sections—one for

---

16. The hertz (Hz) is literally "per second." Therefore, fifty hertz means fifty times per second.

each phase. This also means that the two phases should be distributed more or less evenly in buildings. Being serviced by two phases provides some benefits. Not only does the transformer outside step the voltage down to 120 volts, but it also changes the phase of the currents so that the two currents that service a building are at 180 degrees out of phase from one another. As a result, when both phases are used at the same time (as opposed to one phase with the neutral line), the voltage is doubled to 240 volts. If you live in North America, go check your electrical panel; you will see that some appliances, such as the air conditioning system, the water heater, and sometimes the oven, have a double breaker.[17] These appliances use both phases. Because they require a lot of power, it is preferable to run them on 240 volts rather than 120 volts.[18]

Nowadays, the power grid extends over enormous areas, covering not only regions but also continents. In the United States, there are three large, independent grids: the eastern grid, the western grid, and the Texas grid.[19] Much of continental Europe is connected to a single grid.

Having larger grids is desirable because it is easier to manage changing loads—bigger usually means more stability in the power grid. But when a problem occurs, it can be catastrophic. On August 14, 2003, a large part of the Northeast in North America suffered from a massive blackout. It was a hot day and air conditioners were blasting their cool air, drawing an excessive amount of electricity from the grid, causing a power line to overheat and fail. And when a power line fails, the current is redistributed to other power lines that in turn can overheat and fail. This is called *cascading failure*. Normally, alarm systems are put in place to cut power and isolate smaller areas rather than having entire regions affected, but they malfunctioned in this instance. After this event, a better system was put in place that uses the telecommunication network that we will

---

17. Depending on whether the appliances use electricity to generate the heat or whether they burn gas.

18. This is because, for a fixed power demand, the current is higher for a voltage of 120 volts than 240 volts. And again, current is like flow rate, and high flow rates mean higher losses. If too high a current is drawn, wires might heat up too much and represent a fire hazard.

19. The Lone Star State deserves its nickname when it comes to electricity.

learn about later. Information in fiber-optic cables travels faster than electricity. It is therefore possible to detect a failure and transmit information to the next relay before the current surges.

In part to become more resilient, this enormous grid that we have built will need to be adapted. A quiet revolution in the electricity world has been happening at least since the 1980s, but it has gotten even more powerful since the 2010s.[20] Fewer and fewer people want to rely only on the grid to get their power, and others want to produce their own power in more environmentally friendly ways. As a result, many customers have been generating their own power—for example, by installing solar panels on their roofs. When they need the electricity they generate, they consume it, and when they do not, they sell it to the grid. The principle sounds easy. It is also manageable for a utility when few people do it. The problem comes when many people start to do the same thing. By design, the grid was built to have few large power plants producing electricity and many customers consuming it. That is a one-to-many design pattern. Now, customers want to be both customers and suppliers—a many-to-many design pattern. This arrangement requires a different grid than the one we have. It is as if many of us treated our own water and asked our water utility to buy it from us, but that is not how the water distribution system was built.

Considering the size of and current challenges with the power grid, I am reminded of a quote from the Roman emperor Tiberius: "We have the wolf by the ear and we can neither hold him nor safely let him go." The power grid is a giant beast built by us. We have made our lifestyles dependent on its constant operation—dependent on having billions of gigawatts of electricity traveling seamlessly through a giant network of transmission and distribution lines. I am not sure this is a smart idea. What is certain is that the grid of the end of the twenty-first century will not look like the grid of the beginning of the century. The massive rollout of smart meters (to monitor your electricity consumption in real time),

---

20. A great book on the topic is G. Bakke, *The Grid: The Fraying Wires between Americans and Our Energy Future* (New York: Bloomsbury, 2017).

the emergence of microgrids (neighborhood-scale power grids), advances in electricity storage (i.e., VPP), and the development of demand response strategies (discussed in the previous chapter) are contributing to a new and better grid.

In contrast to electricity, another infrastructure sector is not changing as rapidly. In fact, I would argue that it is doomed to disappear in the not-so-distant future. It is the natural gas collection and distribution infrastructure. We will learn about this subject by traveling to South America. Let us head to Buenos Aires.

POWER DISTRIBUTION IN AN ARCHIPELAGO

Picture yourself in an archipelago (a group of islands). Your goal is to provide electricity to all the islands of the archipelago. How do you do it?

The most obvious strategy might be to build power plants on every island. That way, all islands would be independent. Moreover, islands with more people would have more or larger power plants, and islands with fewer people would have fewer or smaller power plants. This strategy is fine until we think about the energy sources that we can harvest to generate electricity. What if one island had a large river over which a hydroelectric power plant could be built that could supply enough electricity for the entire archipelago? As we have seen, transmission lines can be built to transport electricity. To connect islands, undersea transmission lines can work very well, too.

The archipelago at the center of this story is the Philippines, where I spent my honeymoon in 2014. The Philippines is made up of 7,641 islands. All the transmission network of the archipelago is developed, operated, and maintained by the National Grid Corporation of the Philippines (NGCP). The country has three main power grids. The Luzon grid transmits electricity to the northern part of the archipelago—mainly to Luzon Island, which includes Metro Manila (the capital of the Philippines). The Mindanao grid transmits electricity to the southern islands of the archipelago—mainly to Mindanao Island. In between, the Visayas grid transmits electricity to several islands, including Panay, Negros, Cebu, Bohol, Leyte, and Samar.

During my honeymoon, I spent several days on Bohol Island. Bohol is gorgeous and popular for several things, including its chocolate hills and its tarsiers. It is home to about 1.4 million people. The island is fairly large, with a size of 4,821 square kilometers (1,821 square miles) and 261 kilometers (162 miles) of coastline.

As an urban engineer, I remember my trip well because I went in July 2014. The previous year, on October 15, 2013, the island had been hit by an M_w 7.2 earthquake. Many buildings and roads had been destroyed, but my local tour guide told me that apart from being out of power for three days, they were largely not affected. Three weeks later, however, in November 2013, the Philippines were hit by Super Typhoon Yolanda. Luckily, Bohol was spared from the typhoon. The neighboring island of Leyte to the east was not so lucky. The damages from the typhoon were catastrophic. Millions of people became homeless. Many power plants were damaged, and the Leyte grid was badly affected and inoperable in many parts of the island.

At the time, the peak demand in Bohol was around 62 megawatts,[21] but the power plants on the island could only produce up to 21 megawatts. The rest came from one single 138,000-volt undersea cable from Leyte with a power capacity of 90 megawatts. With the typhoon, the connection was ruptured. As a result, the power in Bohol was out for more than three weeks, which to my tour guide was much worse than the three-day outage from the earthquake. I was amazed by this story. Here was one place that suffered from two disasters in a short time. The first disaster hit the place directly while the second did not, and yet the impacts of the second were worse than the impacts of the first. That story alone speaks of the central place that infrastructure holds in our lives.

By 2022, the peak demand in Bohol had increased to 97 megawatts, and it is forecasted to increase to 377 megawatts by 2040. New electricity generation capacity was built on the island and could produce about 50 megawatts of power (mostly from diesel plants). The rest still came from Leyte, which hosts one of the largest geothermal power plants in the world. Yet since the typhoon, many recognized the fact that relying on a single undersea cable was not conducive to resilience. As a result, in 2024 a new undersea cable was installed between Bohol and another neighboring island, Cebu, which is to the west of Bohol. This 230,000-volt cable has a power capacity of 600 megawatts. Unfortunately, electricity in Cebu is generated mostly from coal-fired power plants, but Cebu is itself connected to the island of Negros, farther to the west, which generates much of its electricity from geothermal, biomass, and solar power plants. In Bohol, plans are

---

21. The technical information for this story was collected from NGCP, "Transmission Development Plan 2023–2040," National Grid Corporation of the Philippines, 2023, available at https://ngcp.ph/Attachment-Uploads/TDP%202023-2040%20Consultation%20Report-2023-06-15-07 -54-06.pdf (accessed May 29, 2024); BEDAG "Bohol Island Power Development Plan," Technical Working Group (TWG) of the Bohol Energy Development Advisory Group (BEDAG), 2017, available at https://ppdo.bohol.gov.ph/ppdofiles/BEDAG/BIPDP_2016-2045.pdf (accessed May 29, 2024); and PH DOE, "List of Existing Power Plants as of August 30, 2023," Republic of the Philippines Department of Energy, 2023, available at https://www.doe.gov.ph/list-existing-power -plants (accessed May 29, 2024).

in place to upgrade the undersea cable between Leyte and Bohol to 230,000 volts as well.

In the future, I could see buildings in Bohol being systematically equipped with solar panels to generate more electricity locally. More storage facilities and devices can also be installed to build a VPP and to be self-sufficient for some time if or when severe extreme events affect the transmission lines between Bohol and its neighboring islands. Yet making Bohol completely independent does not make sense. The island is small and does not have many natural resources that could provide a lot of energy to produce electricity. These connections between islands are essential. Plus, we are always stronger together, and this idea applies to infrastructure, too.

CHAPTER TWELVE

Natural Gas

A Dead Cow, an Oracle, and a Rotten Egg Smell in Buenos Aires

TICK, TICK, TICK, TICK, SWOOSH ... THE FAMILIAR SOUND OF A GAS burner turning on. Many households in the world use gas for space heating, water heating, and cooking. Some households even have clothes dryers that run on gas. Whether they use a gas tank or have a direct gas line to their home, those molecules of gas that they burn have had an amazing journey. To learn about natural gas infrastructure, we will travel to South America and land in the Argentinian capital: Buenos Aires.

I have heard many wonderful things about Buenos Aires, particularly about its European feel that justifies its nickname as the Paris of South America. I would also like to visit Buenos Aires because of its street network, which follows a grid structure unlike any other city in the world. While most intersections in cities are T-shaped (i.e., one street ends in the middle of a segment of another street), Buenos Aires has mostly plus sign–shaped intersections (i.e., where two streets cross each other). In Buenos Aires, no matter the bends and turns, you almost always end up at an intersection between two streets. When I tell that to people, everyone is surprised. They think all cities are like Buenos Aires, with plus-shaped intersections, but it is not the case; T-shaped intersections are much more common. Buenos Aires is unique.[1]

---

1. In a scientific article, my group looked at the street network of the eighty most populated cities in the world. Buenos Aires came as a surprise. Learn more in M. Badhrudeen, S. Derrible, T. Verma,

Thinking about Argentina more generally, I have wanted to visit Patagonia for a long time, and, like most, I am puzzled by how fast the glaciers of Patagonia are melting—partly due to the burning of natural gas. The foodie in me also makes me picture delicious asado (barbecued meat). And after doing research for this chapter, a new facet of Argentina has cropped up in my mind. It turns out that Argentina is home to the second-largest shale gas reserve in the world,[2] and the main area that houses this shale gas has the perfect name considering we are talking about a toxic and flammable gas. The area is called Vaca Muerta, which is Spanish for "dead cow."

Here is something else I did not know: Argentina is the eighth-largest country in the world. From the subtropical north to the polar south, the country spans close to 4,000 kilometers (2,500 miles). In terms of surface area, it is roughly equivalent to the size of the United States east of the Mississippi River. The Vaca Muerta formation is located in the Neuquén Basin, which is geographically in the center of the country, a two-hour plane ride from Buenos Aires. Beyond the Vaca Muerta formation, Argentina also has other large shale gas formations, including the Austral Basin in the south and the Noroeste Basin in the northeast.

The exploitation and transport of natural gas has a tumultuous history in Argentina, filled with nationalizations and privatizations. To give you a flavor for it, as of this writing, the largest extractor of natural gas in Argentina is the company Yacimientos Petroliferos Fiscales (YPF), which was renationalized in 2012 after having been privatized in 1999. The largest transporter of gas is the company Transportadora de Gas del Sur (TGS),[3] which was formed in 1992 after the national company Gas del Estado was privatized. The privatization of Gas del Estado also led to the creation of nine gas distributors across the country, the largest being

---

A. Kermanshah, and A. Furno, "A Geometric Classification of World Urban Road Networks," *Urban Science* 6, no. 1 (2022): 11.

2. The largest shale gas reserve is in China.

3. In English, the name of the company is Transporter of Gas of the South. As you may expect, the company Transportadora de Gas del Norte (TGN, Transporter of Gas of the North) also exists. TGS transports roughly two-thirds of the natural gas in Argentina and TGN the remaining one-third.

MetroGas, which serves the central area of Buenos Aires, commonly called CABA (for Ciudad Autónoma de Buenos Aires),[4] and some of the surrounding districts that are part of the province of Buenos Aires. While MetroGas is a private company, it was bought by YPF, the renationalized company, in 2012. But all companies are regulated by the Ente Nacional Regulador del Gas (ENARGAS), which literally translates to "national gas regulating entity" in English. I did write that the history of natural gas in Argentina was tumultuous.

But enough with the confusion of who owns what and who does what. Let us focus on natural gas itself and on the processes needed to get that "swoosh" when we turn the gas knob. Starting with the basics, we first need to know that, like oil, natural gas is formed over millions of years by the decomposition and compression of organic matter. In fact, natural gas and oil can be present in the same well, with the heavier oil settling at the bottom and the natural gas rising to the top, trapped by a solid layer of nearly impermeable rock. Other wells can be made entirely of natural gas (whether in the form of a gas or a mixture of gas and liquid). Because gas is lighter than air, it also naturally rises to the Earth's surface, which is why solid layers of rocks are needed to contain the gas. Yet cracks in the soil can be present, and some of it can escape in the air. Moreover, because it is highly flammable, these small gas flows emerging from cracks can sometimes catch fire—for example, if ignited by lightning. For the longest time, human populations considered these "forever" burning flames a sign of divinity. In ancient Greece around 1000 BCE, a "burning spring" was found and a temple was built around the flame. A priestess used the flames to make predictions about the future. Her name: the Oracle of Delphi. It was not until around 400 BCE that people in China sought to make use of natural gas, using bamboo shoots to make small pipelines to transport the gas.

The identification of methane proper had to wait until 1776, when Alessandro Volta, the Italian physicist and chemist whose name was used

---

4. Autonomous City of Buenos Aires in English.

as a unit of electric potential (i.e., the volt), studied the chemical properties of natural gas.[5] We had to wait another forty-five years, however, to see any form of industrial use of natural gas with the digging of the first gas well in Fredonia, New York, in 1821, which used the extracted gas to light the city. Even then, transporting gas over long distances remained a problem. Most often, gas was seen as a byproduct of oil that had to either find a market locally or be flared—that is, burned in the atmosphere since the resulting gas is nonflammable carbon dioxide (plus, now we know that methane has a much nastier impact on the environment than carbon dioxide[6]). It was not until the 1920s and 1930s that the long-distance transport of gas was achieved in the United States.[7]

At the global scale, we had to wait until after World War II before natural gas became an important source of energy, once techniques were found to liquify natural gas, reducing its volume to one six-hundredth of the original volume (i.e., by roughly 99.93 percent).

So far, we have discussed natural gas as if it were a single gas, but this is not the case. While every well is different, and while even the composition of a well can evolve over time, the gas that is extracted from wells is composed of both gaseous and liquid elements, such as water, that will need to be removed (sometimes significantly, both in liquid form and as water vapor). The gas itself is made primarily of hydrocarbons and generally dominated by methane (CH_4), but it also contains significant amounts of ethane, propane, and butane, as well as small amounts of other hydrocarbons. Impurities are likewise generally present in the gas, including nitrogen and carbon dioxide, which are removed during various processes described later. The propane and other heavier hydrocarbons are usually removed from the gas because they have higher market value.[8] In the end,

---

5. After reading a paper by Benjamin Franklin that discussed "flammable air."

6. All gases have a global warming potential (GWP) based on their impact once in the atmosphere. Carbon dioxide is given a GWP of 1 compared to methane, which has a GWP between 28 and 36 over 100 years.

7. In 55- to 60-centimeter diameter pipes (22 to 24 inches) operating between 27.5 and 40 bars (400–600 psi).

8. Propane is typically used in grills and portable stoves. Butane is typically used in lighters. Both propane and butane can be used as refrigerants, too.

the natural gas as delivered to your home is mostly made of methane. What we also often forget is that natural gas has no natural odor. The common rotten egg smell of gas comes from an odorant called mercaptan that is added to it purely for safety reasons[9]—so humans can detect gas leaks thanks to their sense of smell only (no fancy technology required).

Once natural gas is extracted from a well, it must be processed to separate its various elements—the extracted substance is termed *wet gas* because it is a mix of gases and some liquids. While some of the processing is at times done right at the wellhead (e.g., to separate some elements that can corrode the pipes), most of it is done in centralized processing plants. The transport system to bring the wet gas from the wellhead to the centralized processing plant is called the *gathering system*, and it usually consists of low-pressure small diameter pipes.

Gas processing at centralized processing plants generally follows four stages. The first stage of processing is to remove all liquids from the wet gas; these include liquid water, as well as crude oil and hydrocarbon condensate, and sometimes even some solids. The removal of liquids can be done by gravity since liquids are heavier than gases, but sometimes wet gas needs to be cooled to favor the condensation of some of the gases that can then be removed by gravity. What we are left with is wet gas without liquids, but it still contains water vapor that needs to be removed.[10] This is the goal of the second stage: dehydration. Dehydration can be achieved primarily in two ways that look and sound awfully similar but are in reality different: absorption and adsorption (notice the shift in the second letter of the word from "b" to "d"). In absorption, water vapor gets dissolved in a liquid such as glycol. In the more popular adsorption, water vapor condenses on solid surfaces such as activated alumina or granular silica. Next, the third stage is to remove other hydrocarbons, such as ethane, butane, and pentane, that have a higher

---

9. This practice has occurred roughly since 1937, after the New London School tragedy in New London, Texas, in which close to three hundred students and forty teachers lost their lives due to an undetected gas leak.

10. Water in pipelines sometimes forms icelike solids that can plug and damage equipment.

market value than methane, which are typically referred to as natural gas liquids (NGLs)—not to be confused with liquid natural gas (LNG), which we will cover later. Perhaps unsurprisingly, absorption and adsorption are used here as well to trap these hydrocarbons, albeit using different agents (e.g., some oil instead of glycol). The final processing stage is to remove the last remaining impurities: carbon dioxide and hydrogen sulfide (i.e., sulfur). Here again, having these gases react with a third, often solid agent (i.e., adsorption) does the trick. In the case of carbon dioxide and hydrogen sulfide, amine absorption technologies are commonly used. That is it. We now have a mostly dehydrated and oil-free gas, called *dry gas*, that consists primarily of methane with some ethane. Newer processing technologies also make use of membranes to separate water and carbon dioxide.

The next step is to send the dry gas over long distances in what is called the *transmission* system, similar to electricity. There are essentially two strategies to transport natural gas over long distances. The first strategy is to cool the dry gas to -160°C (-260°F) so that it becomes a liquid that takes 99.87 percent less space than the original gas in its gaseous form, as we saw above. The liquified form of the gas is called LNG for "liquified natural gas." LNG can then be transported via insulated pipelines to ports where it is loaded onto gas tankers that keep the LNG cool all the way to its destination, where it can be heated and turned back into a gas. Liquifying gas makes sense when it is to be transported over significantly long distances, such as across oceans, but insulating pipes to keep a temperature of -160°C is costly. The second strategy is therefore often preferable, or it can complement the first strategy to transport natural gas from its port of arrival to its final destination.

The second strategy is to use pipelines that transport compressed gas. The main governing force that dictates the direction in which gas will flow is pressure. Gas will systematically travel from high to low pressure; essentially, it is sucked in by the low-pressure area. At the centralized processing plant or at the port of arrival, dry gas is therefore compressed, somewhere between 15 and 90 bar,[11] depending on the system and where it is sent. Large transmission systems generally have pipes with diameters between 0.4 and 1.2 meters (i.e., 16–48 inches). Because of friction losses, gas pressure needs to be boosted roughly every hundred kilometers (i.e., 60 miles), until it reaches its destination.

The amount of gas contained in a given pipe volume is called a *linepack*. A trick used by gas companies is to play with the pipe pressure to increase or decrease the linepack so that, at times, the pipeline itself is used as storage space when it is not needed at the destination. As a result, for the same absolute volume in a pipeline, a higher amount of gas can be stored when the pressure is higher (i.e., when it is compressed). This practice is not possible with liquid oil or water, since they are not practically compressible, but for gas, it works well. When demand is low, the linepack is increased so that the same pipeline contains a higher amount of gas, and when demand picks up, the linepack is decreased and more gas is delivered. This process is called *linepacking*.

To store large quantities of gas, however, linepacking is not enough. Usually, the demand for gas is much higher in the winter since gas is used for space heating,[12] and therefore gas must be stored during the summer months to be used in the winter months. Now, we are talking about enormous quantities of gas. One option would be to build gigantic reservoirs, but containing compressed gas is not easy, and we want to avoid as much leakage as possible. What is the solution? Use natural reservoirs. After all, gas was trapped for millions of years underground with little to no leakage. One option is to use old, depleted gas reservoirs and fill them up with

---

11. The *bar* is a relatively old unit that we saw in footnote 7 on page 154 and in the water distribution chapter in Hong Kong. I would personally favor the use of pascals, but the bar is the industry standard (although the conversion is simple as 1 bar is 100,000 pascals). For those who prefer the imperial British system, 15–90 bars represent roughly 200–1,300 psi.

12. In the United States, data for the 2010s from the Energy Information Administration (EIA) shows that demand for gas has roughly been 40 percent higher in the winter. That said, spikes in demand for gas have appeared in the summer as well, as natural gas has become a major fuel source to generate electricity and electricity consumption in the summer is high to run air conditioners. That is a great example of the vicious circle in which we are trapped. Because gas is burned, the Earth is getting warmer, resulting in a need for more air conditioning, and because electricity is generated with gas, this situation leads to the burning of more gas that will eventually make the Earth even warmer.

gas again (and extract it again when needed). This strategy has become common around the world. Other strategies include utilizing other types of natural underground storage space, such as carefully selected aquifers, salt caverns, and old mines—as long as a layer of impermeable rock is present to trap the gas in place. The advantage of these solutions is that the reservoirs have proven their efficacy already. Moreover, significantly less energy is needed to use them in contrast to building large reservoirs or keeping LNG cool. That said, the likelihood of leaks is not negligible—these reservoirs are natural, and some leakage may occur naturally; also, because the reservoirs were drilled in various places, leakage can occur if these are not sealed properly. This is what happened in Aliso Canyon in California in 2015 when a breach in an old metal pipe occurred at more than 2.6 kilometers (1.6 miles) underground, resulting in the worst single reported natural gas leak in U.S. history.

To constantly monitor transmission lines and storage systems, valves and meters are essential. Meters are used to monitor the system (e.g., the pipe pressure and volume), as well as to measure whether any leakage is occurring. Valves are typically open, but they can be closed to shut off the gas transmission if needed. The monitoring systems are called supervisory control and data acquisition (SCADA) systems. They are present in transmission and storage systems as well as in distribution systems. SCADA systems are also present in electricity and water distribution systems. In fact, we first learned about SCADA systems as they relate to water distribution in Hong Kong. The central nerve of SCADA systems is telecommunication, and telecommunication infrastructure has pervaded virtually every infrastructure since the 1980s, as we will learn when we travel to New York City and San Francisco.

At the end of the transmission system, once the gas arrives at its destination—for example, once it arrives at a local gas utility and before it is to be sent to customers—it must first go through a *gate station*. The goal of a gate station is threefold. First, it is to decrease the pressure of the gas. From 15–90 bar in the transmission pipeline, the gas pressure will be decreased somewhere between 15 bar and 15 mbar (i.e., 0.015 bar), again depending on where it will be sent. Second, it is to measure the volume of gas that is being received and that will be sent in the distribution system. And, finally, it is to add the mercaptan so that the naturally odorless gas now smells like rotten eggs.

At this point, the gas is ready to be distributed and burned. Similar to water distribution, thousands of kilometers of gas pipes are laid beneath city streets. The province of Buenos Aires is serviced by several gas distributors, each having its own geographic territories regulated by ENARGAS. As mentioned, MetroGas is Argentina's largest gas distributor, servicing central Buenos Aires and some of the surrounding districts. By 2020, MetroGas had about 2.5 million customers.

The typical gas pressure delivered to residential customers in Argentina is around 20 mbar (0.3 psi), but not all customers have the same needs. For example, some commercial and industrial customers require more gas. Moreover, as we saw, pressure decreases because of pipe friction. It is therefore preferable to distribute gas at higher pressures through the city and decrease this pressure bit by bit as it arrives at the customer. This practice is equivalent to what is done in the power grid, as we saw in the last chapter, by stepping down voltages bit by bit. In Argentina, ENARGAS differentiates between three types of domestic gas lines based on their pressures: low-pressure lines are rated between 19 and 28 mbars; medium-pressure lines are rated between 0.5 and 4 bars; and high-pressure lines are rated above 4 bars.[13]

So far, we have seen the *bar* only as a unit of measure, but if you look at your gas bill, you will not see any mention of pressure. From a unit perspective, natural gas needs to be measured as a quantity. The easiest quantity to measure is the volume of gas that is delivered, usually in cubic meters (m^3) or cubic feet (ft^3).[14] Look it up if you have a gas bill handy: besides the amount of money owed, you should see a volume of gas consumed. Depending on where you live, you should also see an amount of

---

13. In imperial British units, these values are between 0.28 and 0.40 psi (19 and 28 mbar) for low-pressure lines, between 7.25 and 58 psi (0.5 and 4 bar) for medium-pressure lines, and above 58 psi (4 bar) for high-pressure lines.

14. My gas provider reports volume as "ccf," which stands for "hundred cubic feet," and *therm* (see footnote 17 on page 160).

energy consumed in kWh (kilowatt-hours), kcal (kilocalories),[15] Btu (British thermal unit),[16] or therm.[17]

In Buenos Aires, MetroGas shows a volume in cubic meters. But when we look closely at the bill, we also see the value of 9,300 kcal/m³. The kilocalorie (kcal) is an older unit of energy,[18] and 9,300 kcal corresponds to around 10,810 watt-hours (i.e., 10.81 kWh). To understand it, we must go back one step. The reason why gas is measured as a volume is because gas meters cannot measure energy directly. This situation is a little misleading since it is energy that you are consuming in the end, not volume. Remember that for the same amount of gas, the volume changes depending on the pressure. Moreover, depending on the composition of the natural gas itself (e.g., the proportion of methane versus ethane versus other gases), the amount of energy per unit volume of gas can change as well. In the end, what we seek is a certain amount of energy to convert into heat. Therefore, for the same amount of energy, we need a higher or lower volume of gas depending on its energy content. The story at the end of the chapter will illustrate this point.

That is about it for gas distribution. The elephant in the room, however, is leakage. Even beyond the fact that natural gas is a fossil fuel (thus unsustainable and contributing to climate change[19]), natural gas is also composed of harmful greenhouse gases, and it is toxic and highly flammable. Natural gas leakage is therefore not only harmful to the environment but also a safety hazard. As we learned through our study of water distribution, leakage is inevitable; a pressurized pipe system is bound to

---

15. One calorie is 1.16 watt-hour. One calorie is defined as the energy required to raise the temperature of 1 gram of water by 1 degree Kelvin/Celsius. The term *calorie* comes from the Latin *calor* for heat.

16. One Btu is 0.293 watt-hour. Not unlike the calorie, one Btu is the amount of natural gas needed to heat 1 pound of water by 1 degree Fahrenheit.

17. One therm is 100,000 Btu or 29.3 kWh. Seriously, the use of prefixes in the metric system like *kilo* and *milli* makes everything easier. I cannot understand why some people prefer to keep using the imperial British system. I do not even like to use the unit of *metric ton* for weight anymore; we should simply call it a megagram instead. We use megabytes and gigabytes in telecommunication; why not make the switch for weight?

18. See footnote 15.

19. See the chapter on electricity generation in Cape Town for more information.

leak, and this statement applies to gas pipes as well. A 2018 study[20] esti-mates that 2.3 percent of the U.S. gross gas production alone (i.e., not including distribution) leaks in the atmosphere. A 2015 study[21] estimates a gas leakage rate in the gas distribution system of Boston of about 2.7 percent.[22] Summing the two gives us an estimate of 5 percent, but the real number is likely higher. Moreover, since this book focuses on infra-structure, we will not get into the debate on hydraulic fracturing (better known as *fracking*), but leakage is the main point of debate there as well,[23] as large quantities of natural gas can end up leaking into the atmosphere. Another point we have not discussed is the environmental impact of all the byproducts that are used during the various stages of gas extraction, processing, transmission, and distribution.[24] Furthermore, because it is highly flammable (after all, that is why we use it in the first place), natural gas can also be dangerous.

What is the final take on natural gas? Is it desirable to use natural gas or not? Like most complex questions, the answer is "it depends," and it depends on the alternative. Other fossil fuels (e.g., oil and coal) are known to emit more greenhouse gases when burned. In that respect, natural gas is preferable if fossil fuels are used. Moreover, it is preferable to burn natu-ral gas directly for space heating, water heating, and cooking rather than using it to produce electricity (with one exception, as we will learn in the

20. R. A. Alvarez, D. Zavala-Araiza, D. R. Lyon, D. T. Allen, Z. R. Barkley, A. R. Brandt, K. J. Davis, S. C. Herndon, D. J. Jacob, A. Karion, E. A. Kort, B. K. Lamb, T. Lauvaux, J. D. Maasakkers, A. J. Marchese, M. Omara, S. W. Pacala, J. Peischl, A. L. Robinson, P. B. Shepson, C. Sweeney, A. Townsend-Small, S. C. Wofsy, and S. P. Hamburg, "Assessment of Methane Emissions from the U.S. Oil and Gas Supply Chain," *Science* 361, no. 186 (2018).

21. K. McKain, A. Down, S. M. Raciti, J. Budney, L. R. Hutyra, C. Floerchinger, S. C. Herndon, T. Nehrkorn, M. S. Zahniser, R. B. Jackson, N. Phillips, and S. C. Wofsy, "Methane Emissions from Natural Gas Infrastructure and Use in the Urban Region of Boston, Massachusetts," *Proceedings of the National Academy of Science* 112, no. 1941 (2015).

22. Quite cleverly, both studies used airplanes to measure concentrations of methane in the atmo-sphere, above gas production facilities in the first case and above Boston in the second. But in the future it will be possible to use satellites to measure methane concentration (and thus leakages) in the air in real time.

23. Not to mention the creation of intentional small earthquakes with hydraulic fracturing that have been linked to larger earthquakes later on.

24. The movie *Erin Brockovich* (based on the story of Erin Brockovich) tells the story of a com-munity whose groundwater was contaminated with hexavalent chromium that was used at a gas compressor station to fight corrosion in a gas transmission pipeline.

Brussels story coming up).[25] In the end, natural gas must be phased out. Because it is available on Earth in finite quantities, its use is not sustainable. Moreover, natural gas leaks and emissions related to the burning of natural gas result in greenhouse gas emissions that contribute to climate change. In the end, the cons far outweigh the pros.

From the consumer's perspective, whether we utilize natural gas should make little to no difference, since virtually all uses of gas can be switched to electricity. The end goal is to solely generate electricity from renewable sources. That said, we may still produce natural gas in the longer term; yes, the verb here is *produce* as opposed to *extract* because we can produce natural gas from solid waste. Solid waste and solid waste management are unsung but incredibly important parts of our infrastructure. To see some innovative and ingenious methods of dealing with waste, we leave South America and travel east to the capital of the land of the rising sun: Tokyo.

A TALE OF TWO GASES

Brussels is famous for many things. Besides being a haven for Belgian fries[26] and beer, anyone who has visited Brussels is familiar with the Grand Place, featuring the asymmetrical medieval building that houses the city hall;[27] the Berlaymont building[28] that houses the European Commission; the Atomium with its modern architecture made of nine large atoms; and, of course, Manneken Pis.

Brussels, and Belgium in general, has an amazing history. The country itself is divided in two. In the north, Flanders houses 60 percent of the population, and the main language is Dutch. In the south, Wallonia houses nearly all of the remaining 40 percent of the population, and the main language is French; a small German-speaking community is also present in the eastern end of Wallonia. As

---

25. Furnaces and water heaters have efficiencies between 0.8 and 0.95 compared to natural gas power plants, which have efficiencies around 0.42. But heat pumps are preferable, regardless of how electricity is generated (see the story at the end of the chapter).

26. Yes, French fries are originally Belgian, not French.

27. The legend is that the architect who built the building jumped from the belfry and died by suicide after realizing the building was not symmetrical. But that is just a legend.

28. The large X-shaped building that we see in the news whenever the European Commission is mentioned.

anyone might expect, the two regions have not always been at peace with each other. And what about Brussels? Brussels is special. It is part of neither Flanders nor Wallonia, and it is officially bilingual.

Brussels is not necessarily famous for its gas distribution infrastructure, and yet Belgium went through an important transformation in the early 2020s. In fact, I was initially thinking of featuring Brussels for the natural gas chapter in this book, both because of this transformation and because of a gas explosion that happened in the town of Ghislenghien in Wallonia in 2004. Ghislenghien is located about forty kilometers southwest of Brussels. On July 30, 2004, 24 people died and another 150 were hospitalized after a gas transmission line ruptured, resulting in an explosion—a tragic reminder of the dangers of natural gas systems.

In addition to Belgium being divided into two regions, the country's gas distribution system was divided into two gas types: rich and lean gas. But in the early 2020s, about 1.6 million customers[29] (around half of the country) had to convert their gas system. The Groningen gas field in the Netherlands that had provided lean gas since the 1960s was expected to close by the end of the 2020s because of safety risks—the number of induced earthquakes became too dangerous. As a result, all customers had to switch to rich gas that comes from several countries, including Norway, the United Kingdom, Qatar, and Russia, which is, when you think of it, also a strategy to outsource the safety risks of collecting gas to other communities.

What is the difference between lean gas and rich gas? We must first remember that the pressure and composition of natural gas can change, and therefore its energy content per unit volume can change, too. In the end, the main quantity of gas that is consumed is not a volume but an energy. Therefore, for the same volume of gas, if one type has a higher energy content, you need less of it. While gas meters measure a volume consumed, gas utility companies know the energy content of the gas they are distributing. This is the difference between rich and lean gas in Brussels.

In Brussels, gas wells and gas mixes were different. Because the Groningen gas field in the Netherlands was closing, everyone would now be supplied by the same rich gas. In this case, the adjective *rich* refers to the fact that the gas has more energy per unit volume. In Belgium, lean gas had an energy rating of 10.3 kWh per cubic meter versus 11.4 kWh per cubic meter for rich gas. While half of the country was already supplied with rich gas, the other half had not made the transition until the early 2020s. Compared to lean gas, this means that for the same amount of energy in the gas (i.e., for the same amount of heat generated), a lower volume of rich gas is needed, which may require some equipment upgrade for those switching to rich gas.

---

29. This number includes both residential and nonresidential customers.

While this is a major transition, most consumers are not impacted. Several websites were developed to provide information to the public, including www.gaschanges.be.[30] From this website, I learned that most appliances made after 1978 should be fine and not require any upgrade. Yet many consumers may still opt to upgrade their furnace, not with a newer gas furnace but with a heat pump that does not require gas at all.

Most people have heat pumps at home already. Refrigerators are heat pumps. The way they work is that they displace heat from the inside of the refrigerator to the outside. In other words, refrigerators do not cool the inside but remove heat from the inside, which is why the back of a refrigerator is usually warm (that is the heat that was removed from the inside). Air conditioners are similar. Even if the outdoor temperature is higher than the indoor temperature, air conditioners remove heat from the indoor and displace it outdoors. Air-source and ground-source heat pumps can do the same thing for both heating and cooling. As of this writing, the Russia-Ukraine war disrupted the global gas supply. As a result, many people (especially in Europe) decided to get rid of their gas furnace and replace it with an electrical heat pump.

Eventually, nearly every building will be equipped with heat pumps. They are much more efficient than gas furnaces or even electric radiators. Essentially, gas furnaces and electric radiators convert one type of energy (i.e., chemical or electrical) into another type of energy (i.e., thermal); the best efficiency possible is 100 percent. In contrast, heat pumps displace thermal energy that already exists, which means that with one unit of energy, they can displace several units of heat. Heat pumps have efficiencies between 300 and 700 percent, or even higher. It is only a matter of time before nearly all buildings are equipped with heat pumps.

---

30. Accessed May 29, 2024.

CHAPTER THIRTEEN

Solid Waste Management

From a Garbage War to a Sea Forest in Tokyo

How much trash do you generate? How many times a day do you lift that lid, press that pedal, or open that cabinet door to throw away trash? More generally, how many times a day do you throw something away at home, at work, at school, outside? How many times a week do you take your trash out? And at some point, you know that cool-looking people[1] standing on a step at the back of a garbage truck will come by and take it away, but do you know where? Also, if you had to guess the weight of all the trash that you generate, what would it be? Seriously, take a moment, try to come up with a number . . . what if I told you that you generate your body weight in trash every single month? In some countries, this is the case. However, some countries and cities are better than others. Managing all the trash that we generate can be extremely challenging. Tokyo decided to build an island out of its trash and turn it into a public park.

With nearly forty million people, Tokyo is the most populous city in the world as of this writing. With such a population, it has developed marvelous infrastructure. In fact, the city could have been selected for pretty much any infrastructure system featured in this book. As we saw

---

1. Don't you think they look cool? When I was a child, it was my dream to stand on that step at the back of the garbage truck, hold on to the rail, and cruise along with the crew—a dream I sadly never got to realize.

earlier, how Tokyo has historically dealt with sanitary wastewater and stormwater is fascinating. Moreover, Tokyo has arguably the best urban rail transit system in the world, which combines both an extensive heavy rail system (managed by two companies: the Tokyo Metro and the Toei Subway) and an extensive regional rail system (predominantly operated by the East Japan Railways company). Tokyo is also home to one of the most famous street crossings in the world: the Shibuya Crossing, which I (like most tourists) cross multiple times in awe each time I visit Tokyo. Tokyo's approach to solid waste management is unsung but similarly ingenious. Solid waste management is a captivating infrastructure system that is underappreciated. You may not see it now, but I think you will agree with me by the end of the chapter.

To most people, solid waste management appears easy: Pick up the trash, recycle it if possible, and discard the rest. But the reality is much more complicated. The first thing to realize is that solid waste management is often regulated at the national/federal level with strict laws to limit pollution, but the actual solid waste management is left to the purview of local governments like municipalities. This fact may appear unassuming at first, but we are talking about solid waste, and nobody wants solid waste. Municipalities generally want to get rid of the waste outside of their boundary—out of sight, out of mind—but then it becomes the problem of another municipality. This is why intricate regional plans need to be elaborated or neighboring local governments get into judicial fights. Tokyo experienced this problem firsthand in the 1970s in what is known as the *Garbage War*. As the population increased in the 1950s and 1960s, and as solid waste generation rates increased nearly sevenfold in twenty years, solid waste management became an important issue. Initially, the goal was to make every district in the city self-sufficient so that all solid waste could be handled and disposed of within the districts, but eventually integrated regional plans were developed.[2]

---

2. I learned about this information in H. Kurishima, "Analysis of Current Municipal Solid Waste Management in Tokyo and Future Prospects," in *Tokyo as a Global City: International Perspectives in Geography*, vol. 8, edited by T. Kikuchi and T. Sugai (Singapore: Springer, 2018).

The second thing to realize is that solid waste is mainly a city problem. People generate waste,[3] and as more people settle and live together, managing solid waste becomes an issue. Solid waste management did not exist at the dawn of humanity, but it naturally emerged at the dawn of civilization. That is partly why we know so little about how people lived before cities—archeologists learn a lot by studying old landfills. Traces of solid waste management[4] were found at least as far back as 3000 BCE in the Minoan civilization on the island of Crete (Greece). The roots of modern solid waste management, however, did not start until the nineteenth century, when cities started to grow significantly during the industrial revolution. The creation of modern solid waste management also coincides with the creation of water treatment and distribution systems. The goal of all these new infrastructure systems and services was to promote hygiene in cities.

So far, we have discussed the term *solid waste* without defining it. In Japan, the 1970 Waste Management and Public Cleansing Law[5] defines waste as "refuse, bulky refuse, ashes, sludge, excreta, waste oil, waste acid and alkali, carcasses and other filthy and unnecessary matter, which are in solid or liquid state (excluding radioactive waste and waste polluted by radioactivity)." The definition is long and specific, but we notice two things. First, some types of waste—such as nuclear waste—are not considered solid waste, not because they should not be discarded but because they do not follow the same law. Second, solid waste does not have to be necessarily solid; some liquids can be considered solid waste. In fact, the term *waste* alone is probably a more appropriate term to use, but because it can have different meanings depending on the context (e.g., a "waste of time"), we will use the term *solid waste*. We will avoid commonly

---

3. In fact, the human is the only animal species on Earth to generate waste. No other animal species generates waste like we do.

4. They would dump solid waste in large pits outside the city boundaries and shovel layers of soil at intervals so as not to attract pests. The practice is surprisingly close to what we do nowadays, as we will learn later.

5. Generally called the Waste Disposal Law.

used terms like *garbage, refuse,* and *trash,* but they generally all mean the same thing: *solid waste.*

Following how solid waste is legally defined, we can start to categorize distinct types. Most of us are familiar with municipal solid waste (MSW), which includes most residential, commercial, institutional, and small industrial solid waste. Other types of waste, such as process (or industrial), medical, agricultural, and some residential and commercial (e.g., paint and some batteries), are more hazardous and therefore can have a larger impact on the environment. Usually, they are handled differently from MSW, and they are subject to different laws—hence the importance of properly defining what solid waste is. Finally, we also have construction and demolition waste, which consists primarily of concrete, wood, and other materials and often makes up for the largest proportion of all solid waste (close to 50 percent), most of which can be recycled.

Each category can be broken down further. Some of the main classes of MSW are paper; food; wood; plastics; metals; glass; and rubber, leather, and textiles (i.e., clothes), among others.[6] Knowing the quantity and composition of solid waste is important to determine the right strategies to manage solid waste.[7] Proportionally, food waste tends to be the highest in lower-income countries (around 50 percent). In higher-income countries, paper waste tends to be highest (around 25–30 percent), and food waste tends to be the second highest (around 15–20 percent). That does not mean that people in lower-income countries throw away more food (as we will see in the story featured at the end of this chapter); rather, it means that people in higher-income countries throw away a lot more waste. In fact, there is one variable that significantly correlates with the amount of solid waste a country generates: how rich it is. People in high-income countries almost invariably generate more solid waste than people in

6. Notice that electronics is generally not a category by itself since electronics are made of metals, plastics, glass, and other solid waste types that have their own category.

7. Collecting this data involves not-so-simple solid waste audits. The U.S. Environmental Protection Agency is a good source to learn more about the processes involved; see https://www.epa.gov/smm/instructions-conducting-waste-assessments (accessed August 25, 2024).

low- and middle-income countries. To illustrate this point, we can compare the values of nine kilograms of solid waste per person per month in Vietnam in 2020 with twenty-two in Colombia, thirty-three in South Korea, forty-six in France, and sixty-eight in the United States.[8] After all, capitalism is based on consumption, and mass consumption leads to mass disposal. At the beginning of this chapter, I wrote that you generate your body weight in trash every single month, but it depends on where you live. In Japan, after reaching a peak of thirty-six kilograms of solid waste per person per month in 2001—an already decent figure—emphasizing reduce and reuse strategies in the past twenty years led to a decrease to twenty-seven kilograms per person per month by 2021.[9]

Different solid waste types also have different properties that can dictate how they can be best disposed of. Usually, three types of properties are considered: physical, chemical, and biological. Physical properties involve the size and weight of the solid waste. For example, size can help with sorting; screens can be used to filter large pieces of solid waste from small pieces. Weight is especially considered for transport. Because weight depends on the amount of solid waste transported, a more common unit of measure is weight per unit volume (e.g., kilograms per cubic meter), like the unit used for density,[10] but the term *density* can be misleading since solid waste is made of a mix of materials whose weight is also impacted by whether it is compacted and whether it is dry. Moisture content (i.e., how dry it is) is another important physical property of solid waste. Food waste has a moisture content around 70 percent compared to 4 percent for construction and demolition waste. Typical compacted MSW has a weight around 160 kilograms per cubic meter and a moisture content around 20 percent, but these values can vary significantly.[11]

---

8. S. Kaza, S. Shrikanth, and S. Chaudhary, *More Growth, Less Garbage*, Urban Development Series (Washington, DC: World Bank, 2021).

9. OECD, Municipal waste (indicator), 2024, DOI: 10.1787/89d5679a-en.

10. The density of a material is its weight per unit volume. For example, the density of pure water at 4°C is 1,000 kilograms per cubic meter.

11. An entire table with typical values and ranges is included in my book *Urban Engineering for Sustainability* (Cambridge, MA: MIT Press, 2019).

Chemical properties involve both the chemical composition of solid waste and its energy content. Chemical composition is divided into six components: carbon (C), hydrogen (H), oxygen (O), nitrogen (N), and sulfur (S); the last component is nonvolatile ash (Ash). Identifying the proportion of each component is important to estimate how much energy can be recovered through burning. Namely, a lot of energy can be recovered from solid waste that is rich in carbon; this includes all organic solid waste (organic in terms of chemical composition), such as food waste, paper, plastics, textiles, and wood. Inorganic waste, like glass, metals, and dirt, does not generate much energy when burned.

Biological properties involve looking at the proportion of sugars, fats, and proteins in solid waste. When controlled properly, the decomposition of some organic waste such as food waste and wood by microorganisms can be an abundant source of energy. With the help of air, organic waste can be turned into compost—the technical term is *aerobic composting*[12]—that can then be used as fertilizer to grow crops. When sealed in a tank so that new air cannot enter—called an *anaerobic digester*—the organic waste is still decomposed by microorganisms, but methane is generated that can be burned to generate electricity and/or heat. This method is called *anaerobic digestion*,[13] and this is how we can produce natural gas (as hinted at in the previous chapter). The same process applies to the sludge collected from wastewater treatment plants. Like compost, the *digestate*—the solids left at the end of the process—can be used as fertilizer to grow crops.[14]

So far, we have learned a lot about what solid waste is, but not how to manage it. Solid waste management involves six processes:[15] (1) generation, (2) handling at the source, (3) collection, (4) transport, (5) processing,

---

12. *Aerobic* means *air*. It is the presence of oxygen in the air that helps the microorganisms turn waste into compost.

13. *Anaerobic* means *absence of air*. It is the absence of oxygen during the decomposition process that favors the production of methane.

14. Only if the solid waste used in the first place was controlled so as not to include contaminants such as metals or pharmaceuticals. Otherwise, once the methane has been extracted, it needs to be landfilled.

15. Referred to as *functional elements* in G. Tchobanoglous, H. Theisen, and S. Vigil, *Integrated Solid Waste Management: Engineering Principles and Management Issues* (New York: McGraw-Hill, 1993).

and (6) disposal. Both generation and handling at the source occur in buildings before the solid waste is picked up. Generation (1) refers to the intent to discard something—that is, the intent to turn a previously useful object into solid waste. Handling at the source (2) involves the discarding of the solid waste in a trash can, generally sorting recyclables from nonrecyclables, and the transport to the dumpster or street curb where a garbage truck will pick it up as part of collection (3). Transport (4) involves all transport—for example, from a building to a recycling facility or from a recycling facility to a manufacturing facility or to a landfill. Processing (5) involves all potential processing, including sorting and transformation (e.g., at the recycling facility). Finally, disposal (6) represents the final disposal—for example, in a landfill or in the sea, as is the case in Tokyo. The entire process is therefore much more complicated than simply picking up the trash, recycling it if possible, and discarding the rest. In fact, this is why the entire process is called *integrated solid waste management* when a clear, regional strategy has been put in place.

Not all solid waste management strategies are equal. There is a well-known hierarchy that is famously illustrated as an inverted triangle so that the strategy at the top is the most important. Most people might think recycling is the most important, but think again. Recycling is performed when solid waste has already been generated. The most important strategy is not to generate solid waste in the first place; it is called *source reduction and reuse*—that is why processes 1 and 2 happen in buildings, not after the solid waste has already been collected. Typical examples include reduction in packaging that is destined to be thrown away or reuse of an object (e.g., reusing a shopping bag instead of throwing it away). The second strategy is *recycling and composting*;[16] recycling includes the transformation of waste to give it a new value. The third strategy is *energy recovery*, which generally means burning the waste to generate electricity and/or heat. Finally, the fourth strategy is *treatment and disposal*, which often involves some type of landfilling.

---

16. Composting is itself a form of recycling as we will see later.

This hierarchy is sometimes captured by the four Rs: Reduce, Reuse, Recycle, and Recover. Other versions with more or fewer Rs also exist.[17] Personally, I like to add at least one more R for "Rot" to encapsulate the full solid waste management stream. We can dive into each R sequentially.

We have discussed Reduce and Reuse already; they are the top strategies of solid waste management. Strategies to reduce and reuse range from encouraging people not to buy things they do not need, reusing shopping bags, and using refillable containers to buying secondhand clothes, repairing things that can be repaired (instead of discarding them and buying replacements), and signing up for paperless billing. The Japanese have the word for it: *mottainai*. It expresses a feeling of regret when something valuable is discarded. This term can be translated to "what a waste." Reduce and reuse strategies are in large part responsible for the reduction in solid waste in Japan to twenty-seven kilograms per person per month by 2021.

Next, we have Recycle and Recover, which account for the sorting and processing of solid waste.

The initial sorting of the recyclables versus nonrecyclables is normally done by every individual, by throwing your solid waste in the right trash can. In some countries, recyclables are further divided into subcategories, typically including paper, glass, plastics, metals, and compostable waste (i.e., food and other biodegradable waste). In Japan, sorting solid waste has become notoriously complex. Every city follows different rules. Even within Tokyo, each ward has different rules. Generally, recyclables need to be sorted based on their material type and disposed of in separate see-through bags to be sorted more easily later.[18]

Once picked up by a truck, the recyclables are brought to a recycling facility.[19] Some of the sorting is done by hand and some of it by a machine. One of those machines is an eddy current separator that sorts ferrous metals from nonferrous metals and from nonmetals. It is a fun one to watch

---

17. To my knowledge, the initial form had only three Rs—Reduce, Reuse, Recycle—but I have seen a version with nine Rs: Rethink, Refuse, Reduce, Reuse, Re-gift, Repair, Rent, Recycle, and Rot.
18. It is so complex that it has become a rite of passage for foreigners settling in Japan.
19. The technical term for these facilities is materials recovery facility (MRF).

because nonferrous metals get ejected by the eddy current.[20] Once sorted, the recyclables are packed together, compacted (also called *baled*), and sent to other facilities to be recycled. Formally, *recycling* involves the transformation of waste to give it a new function, compared to reusing, in which the primary function of an object remains.[21] Therefore, composting is a form of recycling since food waste is transformed into fertilizer; technically, anaerobic digestion is a form of both recycling (since the digestate is used as fertilizer) and recovery (since methane is extracted). The largest benefit of recycling is that new material does not have to be extracted from the environment. I will not get into the numbers here, but the recycling of metals, especially aluminum (e.g., aluminum cans), can save a lot of energy and prevent many greenhouse gases from being emitted. Please, do recycle.

Once all the solid waste that can be recycled has been recycled, we move on to Recovery. Perhaps the most obvious form of recovery is through burning. In waste-to-energy processing plants, solid waste can be burned to produce electricity and/or heat. Electricity can be produced and sent to the power grid. Unlike electricity, heat does not travel well over very long distances because of significant heat loss through pipe walls. Nevertheless, it can be used locally to provide heat to nearby areas in *district heating and cooling systems*, as we discussed briefly in Cape Town. Heat can even be used in the summer in adsorption chillers to produce cool air. Remember, however, that not all solid waste can be burned. Only organic waste can be burned (in the chemical component sense, therefore including plastics). Leftover ash and some solid waste such as concrete cannot be burned, but some are still recoverable. Any remaining metals can be extracted from the ash. Some noncombustible solid waste can be turned into new construction materials (e.g., by crushing concrete and

---

20. The way it works is as follows: The solid waste is placed on a conveyor belt, at the end of which an eddy current is created that attracts ferrous metals so they can be sorted out. Nonmetals simply drop in a bin placed right underneath the end of the conveyor belt. Nonferrous metals get ejected by the eddy current so they are thrown away in a third bin placed a little farther from the end of the conveyor belt.

21. Often, I hear people suggesting *recycling* an object while what they really mean is reusing or repurposing it. If there is no transformation, it is not recycling.

reusing it). Then anything left after the recovery process is sent to a land-fill, but it does not have to be sent to a conventional sanitary landfill. Another option is to use the remaining solid waste to make new land.

Through a process called *land reclamation*, many world cities have dis-charged large quantities of nonpolluting solid waste into open bodies of water. Most of the Boston that we know today is built on solid waste, including the affluent Back Bay neighborhood. The famous Ellis Island in New York City that welcomed nearly twelve million immigrants to the United States is mostly made of solid waste (i.e., rubble) that was exca-vated when the New York City subway system was built. And about 20 percent of the entire city-state of Singapore was created through land reclamation.[22]

In Tokyo, land reclamation is a big affair, one started as early as 1592. Like in many cities, land reclamation in Tokyo has historically been a byproduct to solve an altogether different problem. Rubble and dirt exca-vated during the construction of buildings or dredged from the bottom of canals to allow larger ships to navigate had to be disposed of somewhere. Putting it in open-water bodies made sense since new land is created as a byproduct, which can then be used for new purposes. As a result, about 250 square kilometers of land have been reclaimed in the Tokyo Bay— that is over four times the size of Manhattan, 2.5 times the area of central Paris, and the same surface area as the entire archipelago of Saint Pierre and Miquelon, where I grew up. While the process works well with inert materials like dirt, soil, and rubble, MSW is polluted and cannot be dumped in the environment in its raw form. In Tokyo, the solution has been to pulverize all the waste that cannot be burned and to burn all the waste that can be burned to produce electricity and/or heat in waste-to-energy processing plants. These plants are also highly engineered to reduce the pollution from the exhausts (including from mercury and dioxins). Any metals are then collected from the pulverized waste and ash. What we are left with are inert materials that take around 90 percent less space

---

22. Not only using its own solid waste. It also imported a lot of sand, especially from Indonesia.

than the original solid waste and that can be used for land reclamation (technically still called a *landfill*). As a result of the Garbage War, since 1973, a new island has been growing in Tokyo Bay. In 2007, a project was adopted to eventually turn it into a public park. The project is called *Umi no Mori*, which translates to "Sea Forest" or "Groves on the Ocean."[23]

The last strategy, the one at the bottom of the solid waste hierarchy, is landfilling. As the name suggests, landfilling is about the disposal of solid waste on empty land. In the twentieth century, the practice of landfilling evolved from open-air dumps to highly engineered sanitary landfills. There are many types of sanitary landfills depending on the type of solid waste (e.g., hazardous versus nonhazardous), and we can spend some time learning about common landfills that handle nonrecyclable[24] MSW (that was not necessarily burned). The general process is straightforward: dump solid waste on the ground and cover it with dirt at the end of the day so as not to attract pests. This way, layer after layer of solid waste can be piled on top of one another, eventually forming large hills. But can you guess what kind of problems can arise with this formula? Remember that non-burned MSW can include just about any type of nonhazardous solid waste, including food waste. When food waste is thrown away and covered with dirt, it still decomposes, and because of the lack of air, it produces methane, exactly like in anaerobic digesters—that is why the fifth R stands for Rot. Moreover, decomposing organic waste produces a highly toxic liquid called *leachate* that can contaminate groundwater wells below landfills. In modern sanitary landfills, holes are first dug and covered with an impermeable liner, and pipes are installed to collect leachate (to be treated). Gas extraction wells are then installed. Once the landfills are full, another liner is placed on top and all the methane generated is collected—again, this is how natural gas can be produced. Environmental monitors

---

23. To learn more about this process, I recommend the 2017 article "Wasteland: Tokyo Grows on Its Own Trash," published in *Japan Times*, available at https://www.japantimes.co.jp/life/2017/02/18/environment/wasteland-tokyo-grows-trash/ (accessed May 29, 2024).

24. Ideally, landfills receive only nonrecyclable solid waste, but that really depends on how well people sort out their solid waste at home. Too often, significant volumes of recyclable waste end up in sanitary landfills.

are also installed in the area to ensure that leachate and methane do not leak out and pollute the environment.

In the end, some energy can always be recovered, even from landfills. In fact, there is a debate as to whether it is preferable to burn solid waste. On the one hand, burning solid waste enables the recovery of energy, and the resulting ash is easier to dispose of. On the other hand, burning solid waste results in harmful greenhouse gas emissions while sanitary landfills still enable the recovery of some energy (i.e., methane). What is burned also matters: burning trees that had captured carbon dioxide as they grew is different from burning plastics produced from fossil fuels. For the time being, the jury is still out as to which strategy is more environmentally friendly. That said, whichever side people are on, everyone agrees that both burning and landfilling should be the last strategies to adopt, favoring instead the first three Rs: Reduce, Reuse, and Recycle.

Historically, Japan has favored burning its solid waste. After all, the country is not large, and there is not much space for landfilling. Moreover, as mentioned at the beginning of this chapter, no local government wants the solid waste of its neighbor. The easiest solution is to make it disappear in the air by burning it. Burning solid waste is so common in Japan that nonrecyclable waste must also be sorted (burnable versus nonburnable). In 2020, around 80 percent of all the solid waste was burned in Japan (mostly with energy recovery). Only 19 percent of the solid waste was recycled or composted compared to 68 percent in Germany, 60 percent in South Korea, 42 percent in France, and 35 percent in the United States.[25] From what I read, part of the problem is not so much the sorting of waste in Japan but the post-processing of what is considered recyclable. For example, in 2022, while 87 percent of the plastic waste was sorted properly as recyclable, two-thirds of it was burned instead of transformed into new plastic products.[26]

---

25. OECD, "Waste: Municipal Waste," OECD Environment Statistics (database), 2020, DOI: 10.1787/data-00601-en.

26. From K. Matsuoka, "The Japanese Way of Recycling Waste," *Chemical & Engineering News* 102, no. 22 (2024), available at https://cen.acs.org/environment/Japanese-way-recycling-waste/102/i22 (accessed August 25, 2024).

By now, I hope that you are as convinced as I am that solid waste management is as critical an infrastructure system as any. While I cannot predict the future of solid waste management, what I know is that there will only be more reducing, more reusing, and more recycling, so learn how to sort out your trash and start recycling if you do not.

This brings us to the final type of infrastructure that we will cover. It is the last but certainly not the least: in fact, this infrastructure might have become the most widely spread around the world and beyond. It is telecommunication, the very infrastructure that enables us to communicate and exchange remotely. We will start our journey with analog telecommunication in New York City.

THE ZABBALEEN OF CAIRO

As we have seen, managing solid waste is much more complicated than most of us think. Because no one wants to live close to solid waste, and because the disposal of solid waste leads to pollution, what seems to be a trivial process of "pick up, recycle, and discard" ends up being intricate. We have covered the practices adopted in many high-income countries, but not all countries have the resources to implement these practices. And when a country cannot implement practices formally, as we've seen in Bogotá, the informal sector takes over.

Cairo is Egypt's capital city and largest city. It is located just upstream of the river delta of one of the most famous rivers in the world, the Nile. Cairo sits on the eastern shore of the Nile. On the other side of the river lies Giza, Egypt's second-largest city, which is more famous for its monuments: the Great Pyramids and the Sphinx that were built around 2500 BCE. Altogether, Greater Cairo (which includes both Cairo and Giza) is a megacity of more than twenty million people. As we learned, when there are people, there is waste—and twenty million people generate a lot of waste.

Cairo relies largely on the Zabbaleen to manage its waste. To me, the word Zabbaleen sounds elegant. It could be the title of a great song or a poem, but in Arabic Zabbaleen literarily means "garbage people." They number around one hundred thousand people, so about half of one percent of the population. The Zabbaleen people are a minority community, mostly made up of Coptic Christians, whereas Cairo's population is largely Muslim. This not-so-subtle difference makes all the difference, as we will see.

When one aims to manage solid waste, the first thing that needs to be done is to estimate the quantity and composition of the solid waste. Getting data from the informal sector is always tricky. A 2014 World Bank report[27] lists the following composition: organic/food waste: 50–60 percent; paper and cardboard: 10–25 percent; plastics: 3–12 percent; glass: 1–5 percent; metals: 1.5–7 percent; textiles: 1.2–7 percent; others: 11–30 percent. Despite the large uncertainties, we notice the overwhelming presence of food wastes, as is common in lower-income countries. Let us put these numbers in perspective. The World Bank Report[28] used above when listing weight of solid waste by country suggests that Egyptians generate about twenty kilograms of solid waste per month. If 50 percent is food waste, then Egyptians generate ten kilograms per person per month. In comparison, in 2018, the average American generated sixty-eight kilograms of solid waste, and 22 percent of it was estimated to be food waste,[29] giving us fifteen kilograms of food waste per month. The average American therefore wastes more food than the average Egyptian. Regardless, we are talking about enormous quantities of food wastes generated in Cairo.

Considering a regional population of twenty million people, that gives us two hundred kilograms of food waste per month, or 6.7 million kilograms per day. Even if this number is overestimated, we are still talking about very large quantities of food waste. In high-income countries, most of it would ideally be composted/digested and some of it would be landfilled, but the informal sector does not have access to these resources. Another option is to go back to what people have been doing forever: feed it to animals. While not all animals welcome food waste, there is one that welcomes pretty much anything: the pig. That is why the fact that the Zabbaleen are Christians is important. Muslims cannot raise or eat pigs, but Christians can.

The story of the Zabbaleen is famous. Many documentaries have been made about them. Many books and articles have been written about them. On the one hand, they offer a success story. Beyond using pigs to deal with food wastes, recycling rates are also estimated to be very high in Cairo (higher than in Germany) because the Zabbaleen people can process recyclable waste and sell it. On the other hand, they live in insalubrious and unacceptable conditions, as most of the waste is handled in very few areas across Cairo, the largest being Mokattam village, which is better known by its unsurprising nickname: "Garbage City." Medical and other hazardous waste is also often handled with little to no protection. Many children skip school to work with their families, depriving them of an education

---

27. World Bank, *Cairo Municipal Solid Waste Management Project* (2014).

28. Same reference as above: S. Kaza, S. Shrikanth, and S. Chaudhary, *More Growth, Less Garbage*, Urban Development Series (Washington, DC: World Bank, 2021).

29. U.S. Environmental Protection Agency, "Advancing Sustainable Materials Management: 2018 Fact Sheet," (2020), available at https://www.epa.gov/facts-and-figures-about-materials-waste-and -recycling/advancing-sustainable-materials-management (accessed May 29, 2024).

that could offer them a way out of this lifestyle. Their living conditions must improve.

The Cairo government tried to formalize solid waste management in the region by sidelining the Zabbaleen altogether, but it was only partly successful because Cairo has small, narrow alleys where garbage trucks cannot go. Like me, I am sure many readers will have heard of similar stories: minority populations living in unacceptable conditions being sidelined from the one income they have. But we know this is not the solution. While I do not have a solution, I know I will follow the story of the Zabbaleen, and I hope to travel someday to Cairo to meet some of them.

PART FOUR

TELECOMMUNICATION

CHAPTER FOURTEEN

Analog Telecommunication

Can You Hear Me Now in New York City

I STILL REMEMBER THE FIRST TIME I VISITED NEW YORK CITY. MY
girlfriend at the time (who is now my wife[1]) and I landed at John F. Kennedy International Airport and took the subway to check in to our hotel
on the Upper West Side. Then we took the subway again to visit Lower
Manhattan. We exited the subway from one of those small sidewalk exits
with the green railings and colored dots indicating the subway lines serviced by the station and with those stairs that enter the ground as if they
lead into the belly of a beast. I clearly remember walking up the stairways
and being awed by how tall the skyscrapers around me were. The street
was old and narrow and carried little to no traffic. I felt as if I had emerged
in a forest of giant trees made up of concrete and steel buildings. The
skyscrapers enveloped me like a cocoon. I remember being surprised that
the whole scenery challenged my imagination. I had imagined this
moment many times before; I had pictured extremely tall buildings, and
yet what I saw was even bigger than what I had imagined. And in these
buildings, tens of thousands of people were busy working and talking over
the phone, making full use of another type of seamless and silent infrastructure without which New York City would never have come to thrive
as it does: telecommunication.

---

1. The same girlfriend with whom I went to Rome and on a honeymoon in the Philippines.

The last infrastructure system that we will discover is telecommunication,[2] which has transformed society several times, with the fixed telephone (commonly called a *landline*), with radio and television, and, at the turn of the twenty-first century, with the mobile phone and the internet. In this chapter, we will focus on analog telecommunication, which means we will cover only the telephone and the television—or at least how they used to work since the digital revolution transformed everything. Digital telecommunication will be the subject of the next chapter.

Modern telecommunication emerged toward the end of the nineteenth century, more or less at the same time as electricity. Yet, unlike electricity, telecommunication is much older. After all, telecommunication literally means "communication over a distance," and humans have always sought to communicate over a distance. For the longest time, pigeons were used to carry messages. In ancient Greece, people used fire to transmit messages. On the Great Wall of China, beacon fires as well as drumbeats were used to transmit messages. Native Americans used smoke signals to transmit messages. On the island of La Gomera in the Canary Islands, local people used whistling to communicate over ravines. In France, Napoleon used the semaphore—a large wooden articulated structure located on top of buildings—developed by the brothers Chappe to transmit messages across the country in a relatively short time.

The first modern telecommunication system is arguably the telegraph, which used electricity to transmit messages. Most readers will be familiar with Morse code and its beeps—I do not know Morse code, but I always remember that " · · · — — — · · · " (three dots, three dashes, three dots) means SOS.[3] And then came along Alexander Graham Bell, who is credited with inventing the telephone in the years 1875–1876.[4] After Bell, we must acknowledge the work of Guglielmo Marconi, the Italian electrical

---

2. Personally, as an urban engineer, I knew the least about telecommunication as an infrastructure at the outset of writing this book. I am indebted, in part, to Steven Shepard's book *Telecommunications Crash Course* (New York: McGraw-Hill Education, 2014).

3. And do you know what SOS stands for? According to some, it stands for Save Our Souls.

4. I write "credited with" because there is a whole story of how Elisha Gray might have been first (or at least first to create a component that was essential for the telephone).

engineer, for his pioneering work on long-distance radio transmission. Marconi is credited as the inventor of the radio. Finally, by the turn of the twentieth century, humans were able to transmit their voices and ideas over long distances. After all, being able to communicate complex ideas is one of the features that sets humans apart from other animals. It should therefore not be surprising that communicating over a distance (i.e., telecommunication) has always been a central goal of humanity.

Like Tokyo, New York City could have been selected to feature any infrastructure system.[5] But it is New York City's analog telecommunication system that has arguably made it one of the most (if not the most) important cities in the world in the twentieth century. All around the world, billions of people have virtually lived in New York City through the characters in some movie or TV series. It seems to me that Times Square, with its enormous television screens, has become the physical embodiment of what telecommunication can do. And, of course, the iconic Empire State Building, which has been featured in countless movies and must have been photographed billions of times, proudly holds at its very top a broadcasting antenna.

As usual, to learn how it all works, we need to start with some basics. First, we need to realize that telecommunication is all about opening a direct path between two entities. Unlike electricity and water distribution, which tend to link one centralized plant to all its customers, telecommunication is about establishing a direct connection between two customers. In that respect, telecommunication is more like transport since the transport system enables a physical connection between any two addresses.[6] Early phone systems actually offered only direct lines between two phones—a single cable linked two phones, and therefore a phone could

---

5. One interesting story is how New York City used to deal with its solid waste, thanks to a landfill on Staten Island with the ominous name of Fresh Kills.

6. Regardless of the route taken, which can change for every trip, like telecommunication (we will learn about switched networks in the next chapter). In the end, we could say that telecommunication is a type of transport infrastructure, but instead of transporting people, it transports information.

call only one other phone.[7] It is as if private roads connected any two addresses directly. Not only was the system impractical, because you could only call one other person, but the number of cables that were needed was also prohibitively expensive. Quickly, all telephones started to be routed to one central terminal. A person calling would ask an operator to be connected to someone, and the operator would physically place a wire between two jacks so that an electrical current could run uninterrupted, therefore making a direct (albeit temporary) physical connection between two telephones. Later, when phone numbers were invented, the frequency of the dialed numbers could automatically direct the call to a receiver.

In the era of wireless communication, it is worth exploring first how landlines operate, although we will see that the way mobile phones work is similar (the two networks overlap significantly).

First, the act of picking up the phone itself moves a physical switch that lets an electric current flow. This electric current comes from the telephone company. In fact, old landline telephones did not plug into an electrical outlet at all. Telephones still worked during power outages since telephone companies have batteries to ensure service continuity. As telephones started to have more features, they required more power. At some point, they did not need one single infrastructure system to operate properly (i.e., the telephone system) but two (i.e., the telephone system and the electrical grid).

Back to the main story: After the telephone handset has been picked up and an electric current has been established, the caller then dials a number to route the call automatically, with a different beep[8] sounding each time a number is pressed. The phone number itself offers a lot of information. In the United States, a phone number has ten digits, like

---

7. Many of these connecting cables fell during a massive winter storm in New York City in 1888 that led to the burying of all cables.

8. These beeps are a combination of two tones, not one, so that they are not reproducible by a human voice; listen carefully, and you will hear it. The goal is to ensure that a caller gets connected to the receiver they intend to call despite any background noise. Yet clever people found some ways to reproduce certain tones so as not to pay for their long-distance calls.

this: 212-555-1234 (phone numbers cannot start with the digits 0 or 1[9]). The first set of three digits is the area code; in New York City, the original area code is 212.[10] The second set of three digits is the number of the service switching point (SSP), or *switch* for short, that will route the call. And the final set of four digits is the phone number of the receiver—each switch can therefore host a maximum of ten thousand numbers (from 0000 to 9999). If the receiver of the call lives in the same area, the call stays in the local loop (one switch). Otherwise, it must be routed outside, through other switches, until it arrives at the switch of the receiver and to the receiver themselves (hence the term *long-distance* call). All of it can be done relatively seamlessly thanks to computers that route calls properly through the telephone network from a caller to its receiver.

If we turn this system around, we can see a hierarchy. An individual telephone is first wired to a terminal box (a small gray box that welcomes all the phone lines in a neighborhood), which is connected to a switch. A switch is housed in a building to which all the terminal boxes in the area connect, and it is in turn connected to other switches. You can think of it as a large transit system: individual bus stops are connected to a metro station, itself connected to many metro stations that are eventually connected to individual bus stops—this type of network topology tends to be *resilient* because multiple routes are available to connect a call between two telephones.

Before a call is routed, however, we need to ensure that the number exists and that the receiver has not blocked the caller. We also need to determine how to route the call. To do this, there is a parallel telecommunication network called the Signaling System 7 (SS7) that consists of many nodes called signal control points (SCPs). A switch therefore first connects to something called a service transfer point (STP), which queries an SCP as to whether the call can go through and, if so, which route

9. Pressing 0 connected you to an operator. The country code for the United States and Canada is 1. Every country has its own code. For example, Australia is 61, Colombia is 57, Nigeria is 234, Vietnam is 84. Even Saint Pierre and Miquelon, where I am from, has its own code: 508.

10. I write "original" because there are so many customers in New York City that all the possible phone numbers have been distributed, leading to a need for more area codes.

it should take (from predetermined tables). All of these steps are performed in that one or two seconds of silence after the last digit has been dialed and before the phone starts ringing.

Moreover, because voices are transmitted as electrical currents, some losses inevitably occur with distance. As a result, signal repeaters are needed to boost the electrical currents. For analog telephone systems, repeaters are generally simple amplifiers—that is, they amplify the current that they receive. One problem with amplifiers is that they amplify both the voice signal and any added noise, which means that audio quality decreases as distance increases.

For wireless communication, the whole process is similar to landlines. Using some of the science initially developed by Marconi, a mobile phone connects to a cell tower that connects to a switch, and the same process takes place. One big difference is that as people move, their signal can be handed over to other cell towers that may be connected to different switches. That is why a mobile phone tends to be connected to at least three nearby cell towers, trapped inside a triangle,[11] so a signal can be handed over seamlessly. But you know these triangles by their more popular names: cells. That is where the names *cell tower*, *cell phone*, and *cellular network* come from.

Cell tower antennas are often stored in rectangular white boxes, placed vertically and in circles pointing outward, often at 120-degree angles to cover three cells. Each cell tower is planned to cover an area usually abstracted to a hexagon;[12] picture a honeycomb, and at the center of each honeycomb, you have a cell tower. They are often located on top of taller buildings across neighborhoods, but sometimes they are on small telecommunication towers that look a little like mini Eiffel Towers.

---

11. Moreover, using three cell towers enables the geographic localization of the caller, even without a GPS. This process is aptly called *triangulation*.

12. Because it resembles a circle and hexagons tessellate (i.e., they fit into one another without leaving any empty space).

To my knowledge, many water towers in rural areas are used as antennas as well.[13] Naturally, skyscrapers and tall towers are also used for telecommunication. In fact, their antennas are normally larger and used for television broadcasting, too (more on that later). This is the case with the Eiffel Tower in Paris, the Skytree Tower in Tokyo, and the CN Tower in Toronto, whose primary purpose has often been for telecommunication. Similarly, skyscrapers such as the Willis Tower in Chicago, the Burj Khalifa in Dubai, and the Taipei 101 in Taipei are used for telecommunication (as is the Empire State Building, as mentioned before).

Back to the neighborhood antennas (the rectangular white boxes): While they tend to be inconspicuous, once you know they exist, you see them everywhere. In fact, they are there for and because of you—higher-density neighborhoods will have more cell towers. Do you know the series of Verizon commercials with the classic tagline "Can you hear me now?" Telephone companies strategically install cell towers across a region to prevent calls from disconnecting as much as possible, and electromagnetic waves in the air are affected by distance the same way that electrical current is affected by distance in a wire. Some websites and smartphone applications exist to let you find cell towers near you. Personally, I remember being on a boat in Ho Chi Minh City, Vietnam, passing by an informal settlement that had no electricity, but I could see those cell towers every few hundred meters. In some parts of emerging countries, telecommunication infrastructure comes before modern water or electricity infrastructure. In fact, telecommunication infrastructure can help emerging countries leapfrog high-income countries and operate technologies that are not yet widely available (as we will learn about in Oaxaca in the story featured at the end of this chapter). In the twenty-first century, it is possible for nearly every human being on Earth to be connected to the virtual world. This desire to be connected is part of our primal instinct. After all, humans are social animals.

---

13. Once, I remember having poor cell connection on an expressway in a rural area of the United States. I saw a water tower in the distance, and sure enough, as we got closer, my connection came back.

Back to the physical infrastructure: Up to now, we have assumed that every phone line has a dedicated wire that connects it to the switch and to the entire telephone network. For landlines, it should be two dedicated wires since a small electric current carries the information,[14] but those wires are expensive. Ideally, we would like to be able to use two wires to service multiple phone lines. Similarly, for wireless telecommunication, we want individual cell towers to service more than one call at a time. To do so, we need to learn about multiplexing (or muxing).

For analog phones, multiplexing is achieved by playing with frequencies, technically called *frequency-division multiplexing* (FDM). The frequency of human voices ranges roughly from 300 to 3,400 hertz (Hz),[15] but wires can carry a much larger range of frequencies. The entire frequency range on a wire can therefore be divided into subchannels, and voice calls are transmitted on different frequency subchannels so that one single wire can transmit multiple calls in parallel. For example, a phone conversation can be transmitted in the range of 0 to 4,000 Hz, another one from 4,000 to 8,000 Hz, and so on until we have reached the maximum frequency of the wire that still gives us an adequate signal. These ranges are commonly called *bands*—we will get to the term *bandwidth* soon, which you might have heard about. In the end, the full range of available frequencies depends on the type of wire that is used. The equipment used for multiplexing is a subscriber loop carrier (SLC); they are the larger gray telecommunication boxes that you can see sometimes by themselves on the side of the road. A typical SLC is the SLC-96, which can service 96 people over 4 wires (plus a spare if needed) that have 4 wires each; thus, a total of 5 times 4 equals 20 wires for 96 people instead of 192 (i.e., 96 times 2). But we can go even further. Next, after the switch, we can group these lines together and use another type of cable that can

---

14. These two wires are color-coded and twisted; they are called a *twisted pair*. The twisting helps reduce crosstalk and electromagnetic induction.

15. As discussed in the electricity chapter, hertz (Hz) is the unit for frequency, and it simply means "per second." Therefore, human voices send roughly 300–3,400 signals per second.

handle even larger frequencies. Thanks to its full frequency range, the coaxial cable can carry 10,800 calls in one single line.

Coaxial cables have a center conductor that looks like a needle and an outer conductor that forms a cylinder around it. You should recognize it when you see it, and you will likely associate it with your television or internet. As in the telephone, frequencies along a cable are used to transmit analog television, but now the information that needs to be transmitted is more complex. Instead of a *bandwidth* of 4,000 Hz (4 kHz for kilohertz), we now need 6 million Hz (6 MHz for megahertz) since we do not have one audio signal only but two signals in television: one for the video and one for the audio. This is why the increased frequency range available in coaxial cables is needed for television compared to the telephone; otherwise, you would be stuck with only a few channels. And as a video signal is more complicated than an audio signal, the bandwidth required is larger. Note that the *bandwidth* here is the size of the frequency range needed to adequately carry the signal we want to transmit—thus, 6 MHz in the case of analog television versus 4 kHz in the case of analog telephone.

In the absence of a cable, television (as well as radio) is transmitted wirelessly in the air through electromagnetic waves, similar to mobile phone telecommunication. Consider for a moment that the air is akin to a cable with an infinite frequency range, called the *radio spectrum*. Using the radio spectrum properly, formidable amounts of information can be transmitted in the air. From CB radio, maritime and aeronautical communication, and AM/FM radio[16] to television broadcasting, mobile communication, satellite communication, and many more streams of communication, the air around us is continuously filled with a mess of electromagnetic waves that carry information. In the United States, to ensure some order in this invisible mess, the Federal Communications Commission (FCC) allocates frequency bands in the radio spectrum so

---

16. AM stands for amplitude modulation and FM stands for frequency modulation. FM waves are less prone to noise and therefore tend to have better sound quality, but they require a higher bandwidth, and they do not travel as far as AM waves.

that one band is not used for multiple purposes, thereby avoiding disruption in the show you are watching or disconnection of your mobile phone call. The entire spectrum between 9,000 and 275 billion Hz has been allocated—that is, between 9 kHz and 275 GHz (gigahertz).[17] The FCC has an online spectrum dashboard lookup tool to see which frequency bands have been allocated in all regions across the United States. For example, in New York County (Manhattan), the CBS Broadcasting corporation was allocated the 584–590 MHz band[18] and the NBC Telemundo License corporation was allocated the 530–536 MHz band. Moreover, AT&T was allocated many bands, including the 1,885–1,890 MHz band, and Verizon was also allocated many bands, including the 1,890–1,895 MHz band (next to AT&T). The National Telecommunications and Information Administration has published a colorful map that reminds me of a mall map—it is pretty, but it takes a while to find what you are looking for.[19]

But not all frequencies are equal, and there is a simple trade-off. Low-frequency signals are slower, but they can travel farther. High-frequency signals are much faster and can carry much more information, but they cannot travel far because the signal loses energy more rapidly. This is why AM channels are transmitted around 1 MHz and can travel around 160 kilometers[20] (100 miles); there were only a few telecommunication towers at the beginning of the radio. This is also why, largely, AM radio channels are devoted to news and talk radio shows and FM radio channels generally feature music stations.[21] In comparison, 5G technology operates at frequencies over 1 GHz (1,000 times higher than AM radio) but can travel only around 450 meters (1,500 feet). You just cannot have it both

---

17. *Mega* for million and *giga* for billion.

18. Remember that, historically, television channels were given 6 MHz frequency bands. With digital telecommunication, they do not need 6 MHz of bandwidth anymore. We will discuss more in the next chapter on digital telecommunication in San Francisco.

19. I have included a version in chapter 16.

20. Even more at night, up to thousands of kilometers, because the signals can be reflected by the ionosphere, which is why the FCC requires some AM radio stations to stop operating at night.

21. AM radio is cheaper to operate and the sound quality does not need to be great, unlike the transmission of music.

ways. If you want to transmit a lot of information wirelessly, like having a video chat on your mobile phone, you cannot be too far from an antenna.

The fact that higher frequencies do not travel far also partly explains why there are more cell towers in urban areas. Plus, buildings (or solids in general) block signals, which means frequencies travel even shorter distances in dense areas. But there is another reason to have a higher density of cell towers when there are many people. Frequency bands are limited and can carry only so many simultaneous calls. One solution is to have more cells so that the same frequency band can be used more than once over smaller areas. This is another type of multiplexing called spatial multiplexing.

The world of analog telecommunication is fascinating, but it has undergone a revolution. In fact, most of the infrastructure we discussed in this chapter does not operate in an analog world anymore. Everything changed at some point in the latter half of the twentieth century, when we found that it was preferable to divide complex, analog signals into a series of binary signals (i.e., a series of on/off signals). That is when we embarked on the digital revolution. To learn about digital telecommunication, we leave New York City and travel west until we reach the digital capital of the world: San Francisco.

MOBILE TELECOMMUNICATION FOR ALL

Oaxaca is a state in the southern part of Mexico. It is mountainous. The average elevation is around 1,500 meters (5,000 feet) above sea level, giving it an average yearly temperature of a comfortable 18°C (65°F), even though the state is in a tropical region. Oaxaca is home to more than 3.5 million people, most of whom live in small communities and cultivate the land, making agriculture the main economy of the state. Oaxaca is relatively poor, and many people opt to leave.

Added together—mountains, rural populations, low income—we get the perfect recipe to deter utility companies from investing in a region. The reason why is understandable. Utility companies have a business model that favors wealthier urban populations. Installing and maintaining infrastructure is expensive. The larger the area, the more expensive it gets, and small populations translate to small revenues. For telecommunication, mountains add another constraint since electromagnetic waves do not travel well through solids, meaning that more telecommunication towers are needed. This story is typical. Rural areas often get access to infrastructure last, get worse service, and pay more because existing business models are not favorable to them.

In Oaxaca, everything changed in 2013 when the Telecomunicaciones Indígenas Comunitarias (TIC)—or Indigenous Community Telecommunications—network was born. Oaxaca houses one of the largest populations of Indigenous people in Mexico. This element is important, because most municipalities are governed by the system of *usos y costumbres* (customs and traditions) in Oaxaca. In particular, most of the land is communal as opposed to belonging to one single individual. We will see why this factor is important in a moment.

TIC works in a fundamentally different way from regular utility companies. It has adapted its business model to cater to rural populations. First, instead of installing the equipment itself (and having users pay for it indefinitely every month), TIC has the communities themselves invest both time and money in the infrastructure. Off-the-shelf, relatively affordable equipment needs to be purchased to build a local antenna. The antenna is then owned by the community and installed on communal land (therefore free). Second, TIC uses open-source software for everything, from operations to billing, which is free by definition (while private utilities run expensive proprietary software). Third, TIC trains local communities to troubleshoot most hardware problems, either themselves or through a phone call—sending repair people to remote locations can be expensive and can take time. All software problems can be solved remotely from an offsite engineer, whether they are in Oaxaca or in Mexico City.

From an organization viewpoint, this arrangement is made possible thanks to the *usos y costumbres* system, not only for the availability of land on which to install equipment but also for the maintenance of equipment, as one person is generally put in charge of the local network for a number of years and without pay.

From a technology perspective, the system is possible thanks to several technical advances. First, mobile telecommunication requires the sending, receiving, and processing of many signals that used to require mixers, filters, and amplifiers by antennas but can now be done through software. This development means that less equipment is needed (i.e., smaller investments are needed) and that maintenance can be done offsite. This technology is called software-defined radio (SDR); TIC uses GNU Radio, an open-source software. More generally, TIC uses a series of open-source software to route and connect calls and to enable texting. Put together, they call their software solution the Rhizomatica Community Cellular Network (RCCN), which is accessed through another software called Rhizomatica Admin Interface (RAI). Being part of the open-source community and using many open-source software packages, you can find all the codes freely available on the web.[22]

---

22. From the Rhizomatica website, I found that RCCN uses at least the following open-source software: OpenBSC to control the base station (based on GSM technology), Linux Call Router

From a physical infrastructure perspective, the system adopted by TIC follows the three traditional layers. First, local mobile networks are available through a radio access network (RAN) antenna and a base station owned by a local cooperative. TIC was given two radio bands from the Mexican telecommunication commission: the 845–849 MHz and 890–894 MHz bands. The frequency band is typical of GSM technology—GSM stands for Global System for Mobile Communications, and it has been adopted in most countries. These frequency bands are high enough to be able to transmit such complex information as phone calls but low enough to pass through some solids and offer a reasonable coverage area for the community.

The second layer is the backhaul layer. Essentially, a RAN station covers a relatively small area and enables communication within the local community. To connect with other communities, backhaul stations have taller antennas that transmit more complex information at higher frequencies (around 10 GHz, similar to Wi-Fi) from one or several communities. Taller antennas are used because they cover larger areas without encountering a solid—for example, a hill. Moreover, to strengthen the signal and have stations farther apart, antennas point in the direction of other antennas as opposed to transmitting a 360-degree signal (as is the case for RAN antennas to reach everyone around the station). These backhaul stations are normally owned by a small internet service provider.

The third and final layer is a national fiber-optic network owned by an internet service provider. The big advantage to make calls here is that VoIP technology, which we will learn about in the next chapter, can be used to ensure that national calls are as affordable as possible. In fact, in a video on the TIC website, one local says that he now pays one U.S. dollar per month to make unlimited long-distance calls versus having to pay one U.S. dollar per minute before TIC existed.

To summarize how the physical layer works (akin to many systems in the world), information from a mobile phone in a community in Oaxaca is first received by a RAN antenna. If the call is destined for someone locally, it stays in the local RAN station. If the call is destined for someone in the region, it is sent to one or a series of backhaul stations and then to the local RAN station of the person receiving the call. If the call is destined for someone who does not live in the region covered by the backhaul stations, it is sent through a fiber-optic cable as a VoIP call.

One goal of the TIC founders was to ensure that the investment costs could be borne by a local community composed of mostly low-income families. The TIC website states that the initial investment in infrastructure costs around US$10,000, representing around US$50–US$100 per family for a community with 100 or 200 families. In terms of revenue, users pay around US$2.50 per month for unlimited local calls. About 60 percent of the revenues stays in the community to pay for

---

(LCR) to route calls, FreeSWITCH (to route and interconnect popular communication protocols using audio, video, text, and so on), and Kannel (for texting). You can learn more at https://www.rhizomatica.org/ (accessed May 29, 2024).

local costs, including electricity and internet, while the remaining 40 percent goes toward maintaining the telecommunication network.[23]

I hope we will see many more initiatives like TIC in the future. Expecting large companies to change their business models is not the solution. Expecting large companies to be charitable does not seem to be the solution either. Working in partnership with communities and coming up with creative business models appears to offer a much more sustainable and effective solution.

---

23. To learn more about the operations and financial structure of TIC, visit their website: https://www.tic-ac.org/—it has a detailed *Manual de la Telefonia Celular Comunitaria* that has a tremendous amount of information; it is only available in Spanish, but it is easy enough nowadays to find a free translation tool online.

CHAPTER FIFTEEN

Digital Telecommunication

Surfing on the Internet in San Francisco

WHAT COMES TO YOUR MIND WHEN YOU THINK OF SAN FRANCISCO? THE
Golden Gate Bridge? The Transamerica Tower? The Painted Ladies? The
twisted Lombard Street? When I think of San Francisco, it is not an
image but a sound that comes to my mind. Perhaps as a tribute to telecommunication and the transmission of sound, the first line of Scott
McKenzie's 1967 song "San Francisco," with the emphasis on the "an" of
both "San" and "Fran" is stuck in my head: "If you're going to Saann Fraan-
ncisco." As a visual person, I am impressed by the power of this song that
may forever have me associate a city with a sound instead of an image—
come to think of it, I associate Singapore with a smell rather than with an
image, but we will talk about Singapore later. The second line of the song
is "Be sure to wear some flowers in your hair." The line resonates with the
history of the city as a center of the hippie movement. But San Francisco
has changed since the 1960s. From a hippie center, it has become a technology center, though everything started not in the city of San Francisco
itself but fifty to one hundred kilometers south in a region that includes
the cities of Palo Alto, Cupertino, Mountain View, and San Jose, among
others. Throughout the world, this region is better known by its nickname: Silicon Valley.

Silicon Valley takes its name[1] from the fact that computer chips and other computing equipment are silicon based. In the 1960s, while the hippie movement in the San Francisco Bay Area worked toward building a more progressive world through the sharing of ideas, the tech industry was busy building computer chips to change how the world operates altogether.[2] Think of companies like Intel and Hewlett-Packard (HP), two enormous computing equipment manufacturers based in Silicon Valley. Greatly aided by massive waves of immigration[3] throughout its history, the San Francisco Bay Area went from being a medium-sized region on the West Coast of the United States to becoming the center of the virtual world. Then, by the early 2000s, Silicon Valley became known even more so for its software thanks to companies like Apple, Google, and Meta[4]—three giants that live within an ecosystem of thousands of start-ups that fuel them with innovation. No matter their size, all of these companies rely on a complex architecture of cables and satellites that make the physical internet. In fact, it is not a stretch to say that, for these companies, the internet is akin to a virtual blood system, without which they could not survive.

The idiom "last but not least" applies here. Although we learn about it last in this book, the internet has probably become the most important infrastructure system. Even physically, the internet has been invading cities, not only with cell towers but also with cables. I remember being in a meeting about underground infrastructure with some employees from an

---

1. The term was coined by the journalist Don Hoefler, who first used it in an article published in *Electronic News* on January 11, 1971. The article was aptly titled "Silicon Valley USA." Why this region was chosen to become a tech hub is also fascinating. I learned about it on Malcolm Gladwell's podcast *Revisionist History*, in the episode titled "Silicon Valley on the Couch." I will not ruin the surprise by giving the reason in this footnote.

2. Sometimes for the better and sometimes for the worse.

3. The importance of immigration for Silicon Valley is not to be underestimated. I would not be surprised to learn that the vast majority of innovations in Silicon Valley came from people who were not born in the United States. Even the most basic transistor used in all computing equipment—the metal-oxide-semiconductor field-effect transistor (MOSFET)—was invented by two immigrants: Mohamed Atalla from Egypt and Dawon Kahng from Korea. It was also invented not in the Silicon Valley but at Bell Labs (a giant in telecommunication).

4. Two other enormous companies, Amazon and Microsoft, are located in Seattle, not Silicon Valley.

electric utility company. During the meeting they mentioned how "the big four"—water, electricity, gas, and telephone—used to exchange information easily about the location of their infrastructure so they would not cause one another damage when digging up a road. With the internet, everything changed. Empty space under many city streets (especially in central business districts) filled up with millions of kilometers of internet cables from competing telecommunication companies. In addition to the invisible mess of electromagnetic waves in the radio spectrum, telecommunication is responsible for a physical mess under city streets.

But we are getting ahead of ourselves here. Let us go back to the basics. The internet is all about the telecommunication of digital information. But what is digital information?

Digital information is represented by long series of one of two things: on or off. Taking the example of an electric current, all variations matter in analog signals. The values 1.3 volts, 2.5 volts, and 3.7 volts all carry a different meaning and copy the frequency recorded by the voice of a caller in the case of the telephone call. In a digital signal, the voltage can be either 0 or not 0. Therefore, despite some signal loss, whether a voltage is 1.8 volts or 3.7 volts does not matter since it simply means it is not zero.[5] One benefit of digital information is that signal loss is much less of an issue. Fewer repeaters are needed. Plus, signals are easier to repeat. But we are barely scratching the surface as to why digital information is superior.

The unit of measure for a binary signal (i.e., on or off, 0 or not 0) is called the *bit*, for binary digit. The term is attributed to Claude Shannon, recognized as the father of information theory.[6] The bit is elementary and contains little information in itself, but we can combine several bits together to mean something. For example, combining two bits together

---

5. Or, more accurately, it is above a certain threshold such that any current above the threshold is 1 and any current below is 0.

6. Shannon published the seminal article "A Mathematical Theory of Communication" in 1948. The article includes something called *information entropy* as a measure of uncertainty or randomness associated with an event (e.g., the next word in a sentence). It is defined as $H(x) = -\sum p(x) \log p(x)$, where $H(x)$ is the information entropy of event x and $p(x)$ is the probability of event x happening.

gives us four combinations—00, 01, 10, and 11—that mean four different things. In practice, we want to select a number of bits high enough to be able to reproduce the original complex signal. Or, when typing on a computer, we want to make sure that we have a series of bits for "a," another for "b," and so on for all possible characters. In practice, a series of eight bits was selected since it can reproduce a total number of 256 characters; a combination of eight bits is called an *octet*. For example, the octet 01100001 means "a" (lower case) and the octet 01000001 means "A" (upper case). Even then, 256 combinations are not enough to capture all possible characters, and some characters are composed of two octets (such as "à," which is represented by 11000011 10100000).

Another common term for a combination of bits is the *byte*, which can be composed of eight bits as well, but the term has been used for a combination anywhere between 1 and 48 bits. Even though the term *byte* is more common in the United States,[7] we will use the octet because it literally means "set of eight,"[8] and therefore its definition cannot change in the future, unlike the byte.

To digitize a voice into a series of octets, the continuous voice signal is transformed into pulses, and the height of the signal at each pulse is turned into an octet.[9] For example, say we have the original voice data as an electric current with the following values for individual pulses: 1.3, 2.5, and 3.7 volts. Each value is turned into an octet that is transmitted in a wire, and the signal is reconstructed once it arrives at its destination. This is what Voice over Internet Protocol (VoIP) does.[10] By digitizing sounds, VoIP has enabled the transition from expensive to extremely low-cost

7. The unit of space for your hard drive, for example, may be in GB (gigabytes) or TB (terabytes), and, at the time of this writing, a byte is usually defined as a series of eight bits.

8. Think octagon; the polygon with eight sides like the stop sign in traffic.

9. Using a technique called pulse amplitude modulation (PAM). Typically, 8,000 pulses are recorded per second—the frequency is 8,000 Hz—to match approximately twice the frequency of a human voice (around 4,000 Hz, as seen before), to follow something called the Nyquist sampling theorem after the Swedish-born American electrical engineer Harry Nyquist. And because there are 8 bits to an octet, a sampling of 8,000 pulses gives $8 \times 8,000 = 64,000 = 64$ kbits/s (or 8 ko/s, kilo-octets per second), the standard bit rate for digital audio.

10. At the time of this writing, popular VoIP platforms include Skype (often recognized as the company that changed the telephone industry), Telegram, FaceTime, and WhatsApp, to name a few.

international calls since VoIP relies on the internet as opposed to the telephone system. In fact, in a short period of time, VoIP has dramatically transformed the telephone industry all over the world.

The real power of digital information is that it is extremely easy to compute[11] and transmit. Essentially, to transmit digital information, any system that can be turned into a binary system can be used. Beyond using an electrical current, the most famous medium used to carry digital information is light (and the absence of light); light is 1 and no light is 0. The cables that carry light are called *fiber-optic cables*. They are composed of a central glass fiber that carries the light from the beginning to the end of the cable by bouncing off the walls of the fiber. The technology itself (as well as the term *fiber-optic*) was discovered by the Indian American scientist Narinder Singh Kapany in 1953 while he was pursuing his doctorate in the United Kingdom. While light in a vacuum travels at three hundred thousand kilometers per second, in fiber, it travels at about two-thirds of that speed, so around two hundred thousand kilometers per second, which remains extremely fast. To put this number in perspective, the circumference of the Earth is roughly forty thousand kilometers, which means that light in a fiber-optic cable can travel five times around the Earth in one second. To further sweeten the pot, with light, we can also play with wavelength like we play with frequencies in frequency-division multiplexing (FDM)[12] to transmit more information at once; this process is called wavelength-division multiplexing (WDM).

Speaking of multiplexing, FDM can be used in digital telecommunication as well. In particular, wireless frequency bands can be divided into eight small bands so that each bit of an octet can be transmitted over separate bands instead of in series over one single band. In fact, thanks to digital telecommunication, the bandwidth needed to broadcast television channels has been greatly reduced, which means that more channels can

---

11. The metal-oxide-semiconductor field-effect transistor (MOSFET) present in computer chips is made to compute binary values. That is what it does.

12. As seen in the previous chapter on analog telecommunication in New York City.

be transmitted in an existing six megahertz band or that unused frequencies are now available for other purposes.

The remaining bit of information (pun intended) we need to learn about is the infrastructure required to transmit this information. Let us dive into digital telecommunication and venture into the world of the internet. Everything you are sending or receiving through the internet first gets converted into a series of octets before it is sent into a large network that no one controls but everyone loves: the internet. The reason it works so well is thanks to a set of protocols that ensure the quality of the information that is sent/received. The most common representation (or abstraction) used to describe how the telecommunication of a computing system works is the Open Systems Interconnection (OSI) reference model.

The OSI model has seven layers: (1) physical, (2) data link, (3) network, (4) transport, (5) session, (6) presentation, and (7) application. We will go through each layer, starting from the last. To illustrate how it works, let us see what happens when we want to visit a website. We just entered "www.wikipedia.org" in our browser—what happens now?

First, the address "www.wikipedia.org" is called a uniform resource locator (URL). Most readers will know that "www" stands for World Wide Web. The World Wide Web was initially created by Tim Berners-Lee,[13] an English computer scientist, during his time at CERN in 1989. The World Wide Web is the collection of documents that are identified by their URLs and connected by hyperlinks to enable you, when you click them, to hop from page to page and from website to website. The World Wide Web is what people generally refer to as the internet but, technically, the internet is "the global system of interconnected computer networks," according to Wikipedia.[14] In other words, the internet is the physical equipment that enables you to travel through the World Wide Web, not the web pages themselves.

---

13. Who also created the first web browser.
14. https://en.wikipedia.org/wiki/Internet (accessed May 29, 2024).

Moreover, have you ever seen the four letters "http" before the "www" in the URL? Do you know what they stand for? HTTP means hypertext transfer protocol, and we are now entering the world of protocols. Thanks to HTTP, your computer will start to talk seamlessly with the Wikipedia server. It will ask it whether it is okay to see what is at "www.wikipedia .org" and, if so, to send us the information. Sometimes we have five letters: HTTPS, in which the extra "S" stands for *secure*, meaning that the connection is secure. Ensuring that you have the full five letters HTTPS is especially important when you are sending/receiving private information, such as when you are checking your emails and making purchases online. Because Wikipedia has users who log in and add/edit entries, it needs to be secure, and therefore I can see that my browser displays the address "https://www.wikipedia.org/."

Using the web developer tool of my browser, when the page loaded up, I saw that Wikipedia sent me 93 kilo-octets of information. Following the OSI model, here is what happened.

First, the application (7) layer is at the interface between the running software package (e.g., my browser) and the internet. In my case, it converted my query of "https://www.wikipedia.org/" into a message to be sent. On the Wikipedia side, it ensured that the information could be read once it arrived to me by using an agreed-on protocol, which was HTTPS in our case.

Second, the presentation (6) layer added information telling the receiver what to expect, including that the provided information followed the HTTPS protocol. The presentation layer can also encrypt and compress the information to be sent. Encryption is necessary so that the signal is not readable by anyone else who may intercept it. Compression is necessary so that as few bits as possible are transmitted, which both allows us to view our page faster and helps control congestion (more on congestion later).

Third, the session (5) layer manages the communication between two computing systems. Practically, it opened a session between my computer and Wikipedia, and it managed the communication until I closed the page or until Wikipedia and I got disconnected. This way, if I clicked on a link, Wikipedia would know it was me and send me information as opposed to sending it to a random person.

Fourth, the transport (4) layer divided all the information that I sent into small packets of a few octets that can be sent easily over the internet. Similarly, the transport layer of the Wikipedia server broke down the information sent to me into many small packets.[15] My transport layer then ensured that all packets sent by Wikipedia had been received successfully, and, if not, it asked for the missing packets to be sent again. Specifically, I saw that my request was sent over twenty-seven packets and then that I received sixty-four packets from Wikipedia of an average size of 1.45 kilo-octets. There are many reasons for sending small packets instead of a long stream of octets. For starters, if an error occurs on the way to one packet, only that one packet needs to be sent again, not the full message. In fact, ensuring that the full message has been received successfully is the responsibility of the transport layer; if some packets are missing, the transport layer asks for them to be sent again. Moreover, having small packets is a type of multiplexing in itself called time-division multiplexing, so that telecommunication equipment is not monopolized by one user (e.g., so as not to have to wait a long time to load Wikipedia because someone else is streaming a movie).

There are two more concepts that we should discuss about the transport layer before we move on to the next layer. First, the type of transport network that we are using when visiting Wikipedia is said to be a *switched network* because all cables are shared by a lot of people, in contrast to *dedicated networks* that are devoted to specific customers.[16] Second, the transport of individual packets can follow two types of connections.

15. Technically, packets are called protocol data units (PDUs). PDUs are different in each layer and get wrapped around like an onion with more information as they move from layer to layer—for example, to specify which protocol to use and the origin and destination addresses.

16. From a cyber-resilience point of view, dedicated networks are more secure because only information from specific customers who own the network can be transmitted and the transport service is as good as it gets (no congestion problems because of other customers). That said, dedicated networks are also vulnerable because information transmission can stop if cables malfunction or are taken out voluntarily.

In connection-oriented protocols, all packets follow the same route, in an ordered manner, following the first packet that was sent all throughout the session. In contrast, in connectionless protocols, individual packets can follow different routes, attempting to take the fastest route, and all packets are reassembled properly once they arrive at their destination. We will touch on these two types of connections again later.

Fifth, the network (3) layer routes the information being sent/received through the internet and handles congestion. It involves not only my computer and the Wikipedia server but all intermediary routers—called *internet exchange point* (IX or IXP)—where the information might transit. When I sent my request to view Wikipedia, small packets left my router and found their way to the Wikipedia server thanks to its IP address (discussed later), hopping through IXPs. Routing these messages is the responsibility of the network layer, and there are two families of routing protocols: *static* and *dynamic*. Static routing protocols always use the same route to connect two pieces of computing equipment, regardless of the level of congestion. In contrast, dynamic routing protocols select the path that maximizes the transmission speed of the information. Needless to say, dynamic protocols are more commonly used. When I requested to view what was on "www.wikipedia.org," an IXP received my packets, looked at their destination, and sent them to another IXP that they thought would be closer to my destination based on previous experience, and this next IXP did the same thing until my packets reached the Wikipedia server. Using the command "traceroute" on my computer, I saw that my connection to Wikipedia went through nine IXPs. Moreover, my browser showed me that the information from Wikipedia took 256 milliseconds to reach me back for the 93 kilo-octets that were sent but that, once uncompressed, amounted to a total of 177 kilo-octets of information (hence the value of compression done in the presentation layer). Overall, this is quite a feat in about one-quarter of a second.

The network layer is also responsible for dealing with congestion. Several protocols exist to manage congestion, such as rebalancing some of the packets to IXPs that are not as congested or even discarding some packets altogether, which is fine since the transport layer would simply ask for them to be sent again. In the meantime, a congested IXP may tell its neighboring IXPs that it is congested to help rebalance the load thanks to the dynamic routing protocol.

Sixth, the data link (2) layer ensures that all packets have kept their integrity—that is, it ensures that no errors were introduced by the physical layer. It does not look at the full message but at individual packets, one by one, as they come along. And if errors are found, the packets are discarded, and the transport layer will request new copies soon after. To check for message integrity, the bits in the packets are considered as mathematical values instead of octets. Some mathematical calculation[17] is performed, and the value is added to the packet. As the message travels through the internet, the same mathematical calculation is repeated, and if the calculated number matches the stored value, then the message carries on; otherwise, it is discarded. This procedure implies the data link layer knows in an endless series of 0s and 1s where a packet starts and where it ends. For that, another protocol was established so that every packet starts and ends with a *flag* that consists of an octet (i.e., one of the 256 possible octets) that is only ever used as a flag. At the time of this writing, that octet is 01111110. And if this sequence of bits is present in the main message as well (e.g., in the middle of two adjoining octets), a 0 is added before a series of five 1s to change the sequence and prevent it from tricking the data link layer into believing the packet is over. The practice is called *bit stuffing* or *zero-bit insertion*. All extra 0s are then removed when the packet arrives at its destination—that is, when the network layer finds a packet with a 0 followed by five 1s, it removes the 0.

Lastly, the physical (1) layer is the physical computing equipment, such as your router, public Wi-Fi[18] devices, cell towers, IXPs,[19] fiber-optic

---

17. Many error-detection and correction methods exist. Two popular families are checksum and cyclic redundancy check (CRC).
18. Do you even know what Wi-Fi stands for? It stands for Wireless Fidelity. It is a set of protocols, including which frequencies to use, to enable wireless Web access.
19. The first time I learned about them was when I read Andrew Blum's *Tubes: A Journey to the Center of the Internet* (New York: HarperCollins, 2012), which is very accessible and a pleasure to read.

cables, servers, satellites, and so on, that enables digital telecommunication in the first place. In the context of this book, the physical layer is the infrastructure that physically enables digital telecommunication. Like the other types of infrastructure, it is expected to work seamlessly and silently without failing, but the physical layer can (and often does) fail. Nevertheless, because the physical layer is so well distributed around the planet, even if a communication channel fails (e.g., a cable is ruptured or a cell tower is out), it does not take long before packets are sent through other routes. You may remember that we discussed how resilient the telephone system was in the previous chapter thanks to its network topology that allowed a phone call to be routed in different ways if a line was out or congested. It is the same thing with the physical layer. In fact, the physical layer is partly composed of the telephone system itself. Anyone connecting to the internet through a digital subscriber line (DSL) is using the telephone network. Other people may connect through cable, which points to the coaxial cable that is normally used to broadcast television and which tends to be faster.[20] As we can tell, one of the strengths of the internet is its ability to piggyback on already existing infrastructure since all that is needed in the end is the ability to produce a signal that is either on or off (i.e., binary). But thanks to their lightning speed, fiber-optic cables are preferred to transmit digital information. For a long time, fiber-optic cables were used only by telecommunication companies, which would then redistribute digital information on telephone and cable lines, but, as of this writing, they are increasingly being installed directly in people's homes. Colloquially, they are simply known as "fiber." Finally, some people may connect wirelessly as well, either to a nearby router that may connect through a DSL, cable, or fiber-optic cable or to a cell tower that is likely connected to the internet itself by a fiber-optic cable.

In terms of ownership, telecommunication utilities tend to build their own system. That is a point that must be highlighted. With water,

---

20. Because coaxial cables are able to carry higher frequency ranges (as discussed in the previous chapter) as opposed to twisted pairs used by the telephone industry.

electricity, and gas infrastructure, only one system exists. In cities, there is only one power grid. If you have the option to choose your electricity provider, whichever you choose, they share the same distribution infrastructure. With telecommunication, internet service providers (ISPs) tend to build their own systems. As a result, when you change ISP, a technician usually needs to come in to physically disconnect your line from your old provider and connect it with your new provider. That means that millions of kilometers of cables have been installed across the world, often next to one another. Figure 15.1 shows the world submarine cables that keep us connected. All these cables are owned by different entities and offer redundancy if/when they break. In the end, no one single person/authority controls the internet (i.e., the physical layer), which means that no single person or authority can stop all of it at the global scale.

That completes our review of the OSI model, which is, again, just a reference model (an abstraction). As of this writing, the most commonly adopted set of protocols used to navigate the internet is the TCP/IP—called the internet protocol suite—which stands for transmission control protocol (TCP) and internet protocol (IP). The internet protocol suite is abstracted into four layers only: link, internet, transport, and application (to which we could add a physical layer). The application layer more or less mirrors both the application and the presentation layers of the OSI model. The transport layer of TCP/IP more or less covers both the session and the transport layers by ensuring that a session is established between two computers and the quality of the messages sent and received. In terms of connections, the TCP is a connection-oriented protocol—that is, once a session is established and adopts a certain route, the same route is used by all packets until the end of the session. In the future, the TCP protocol is expected to be replaced by a connectionless protocol powered by the User Datagram Protocol (UDP).

The internet layer of TCP/IP mirrors the network layer in the OSI model. To find where packets should be sent, every single piece of computing equipment is given an IP address (usually temporary) composed of

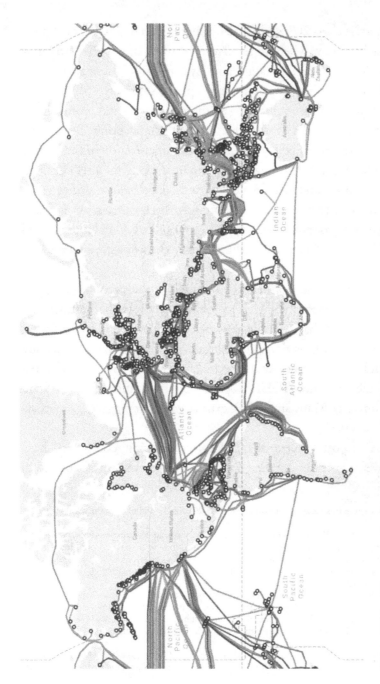

Figure 15.1. *Map of world submarine fiber-optic cables.*[21]
Courtesy of submarinecablemap.com and TeleGeography.

21. The map was collected from https://www.submarinecablemap.com/ (updated May 29, 2024). It was produced by https://www.submarinecablemap .com/ and TeleGeography (https://www2.telegeography.com/) (accessed May 29, 2024).

four numbers, such as 208.80.154.224,[22] that is akin to a telephone number and used to know where to send/receive information. Then, to know where IP addresses are, Domain Name System (DNS) servers are used like signal control points (SCP) in the telephone system.[23]

By now, you may be confused by all these acronyms and protocols. To keep some order in the world, the Internet Engineering Task Force (IETF) of the Internet Society as well as a suborganization of the United Nations called the International Telecommunication Union (ITU) look after all protocols, standards, frequency bands, and so on, and they make recommendations that are generally adopted by countries. As a result, wherever you are around the world, when you connect to the internet and type a URL such as "www.wikipedia.org" in your browser, you are led to the right place.

Back to San Francisco, being the center of the digital world, you would expect that everyone in the region had access to the best internet service possible, but that's not the case. Silicon Valley is a global leader for its software. Apple, Google, and Meta are mostly software companies; they are not ISPs[24] (internet service providers). Popular ISPs in the United States include Astound, AT&T, Cogent Communications, Verizon, and Xfinity (Comcast). Many small ISPs exist as well that piggyback on existing infrastructure.[25]

A 2023 report[26] notes that telecommunication companies have invested over two trillion dollars on wired and wireless infrastructure in

---

22. The day I wrote this paragraph (November 17, 2020), this IP address was the address of www.wikipedia.org.

23. And because your IP number changes—for example, if you are traveling—every piece of computing equipment is given a media access control (MAC) address so the IP addresses know where to send the information they receive. Put differently, the MAC address is like your name and the IP address is like a phone number. You may be in different buildings and be reached at different phone numbers, but your name does not change.

24. Although Google tries to be. It has built its own fiber network.

25. Leasing bandwidth from big providers. Customers usually get the same service (internet speed) for a lower price, but they also receive a lower customer experience. That's how these lower-cost companies remain profitable. That is true for mobile telecommunication too. Smaller players often piggyback on the infrastructure of larger players.

26. USTelecom, *2022 Broadband Capex Report*, USTelecom.org (2023), accessible at https://ustelecom.org/wp-content/uploads/2023/09/2022-Broadband-Capex-Report-final.pdf (accessed May 29, 2024).

the United States between 1996 and 2022. Cell towers, IXPs, fiber-optic cables, servers, satellites, and all other types of physical telecommunication infrastructure that provide us access to the virtual world do not come cheap. Because all this infrastructure is private, it is also very difficult to get data on how much of it there is. I could not find how many kilometers of telephone, cable, and fiber lines are present in San Francisco, but I could find that there were 28 ISPs.[27]

On the website https://www.broadbandmap.ca.gov/,[28] you can see the internet services available at specific addresses. I entered the address of one of the famous Painted Ladies[29] on Steiner Street in San Francisco and found two ISPs: Comcast (Xfinity), offering speeds up to twelve hundred megabits per second (using cable), and AT&T, offering speeds up to twelve megabits per second (using DSL). Because telecommunication infrastructure is expensive, private companies try not to compete excessively with one another. It is rare to have more than two options.[30]

In the twenty-first century, having broadband internet access has almost become a right, like having electricity access in the early twentieth century. Yet not all places are equal. People living in cities tend to have more and better options than people living in rural areas (like we saw in Oaxaca). From the website https://broadbandnow.com[31] I found that, in San Mateo County (where San Francisco is), virtually 100 percent of the population has access to fiber. In rural Modoc County in the northern part of California, only 1.5 percent of the population has access to fiber and only 20 percent has access to speeds above 25 megabits per second. Rural areas are usually worse off, but low Earth orbit (LEO) satellites may help, as we will see in the story at the end of this chapter.

---

27. I am guessing three to five systems owned and operated by large ISPs and more than twenty small ISPs that piggyback on existing infrastructure. Or some providers may be highly specialized and have only a few clients.
28. Accessed May 29, 2024.
29. Famous, colorful row of houses in San Francisco.
30. Where I live in Chicago, I have two options: Astound and Comcast.
31. Accessed May 29, 2024.

Our journey is almost over. We have now learned a lot about the seven infrastructure systems featured in this book: water, wastewater, transport, electricity, natural gas, solid waste, and telecommunication. To wrap up, it is fitting to travel to the city that might have the most advanced infrastructure in the world. Let us go to Singapore.

THE CONSTELLATIONS THAT CONNECT US

An integral part of telecommunication infrastructure is satellites. Since the invention of radio transmission by Guglielmo Marconi, humans have used electromagnetic waves to telecommunicate wirelessly. We have also learned that electromagnetic waves do not travel well through solids. That is why telecommunication towers have been built in cities around the world and why tall skyscrapers are equipped with antennas. The higher an antenna is, the larger the area it can cover. The obvious next step to reach higher altitudes is to go into space.

Since the 1960s, the preferred strategy has been to launch large satellites into orbit at an altitude where they naturally rotate around the Earth at the same speed as the Earth. This happens at an altitude of close to 36,000 kilometers[32] (around 22,200 miles). These satellites are called geostationary (GEO). The main advantage is that a dish pointing at a GEO satellite never has to be repositioned. Everyone who has satellite TV knows this fact—or at least they know that once the dish has been installed, it does not need to be repositioned. Moreover, at such a high altitude, only a few satellites are needed to reach many people. For example, when I wrote this story, the popular satellite TV provider DirecTV operated eleven satellites to serve the United States and parts of Latin America. The main drawback is that communication to GEO satellites is slow because the distance is large. This is referred to as *high latency*. It is okay to watch TV since the communication goes one way, but it is not ideal to have a phone conversation or even less to browse the Web seamlessly since signals are sent back and forth.

---

32. The equation here is $r = \dfrac{G \cdot M_E}{v^2}$, where r is the radius to the center of the Earth, G is the gravitational constant (6.674×10^{-11} m³/[kg·s²]), M_E is the mass of the Earth (5.97219×10^{24} kg), and v is the speed (the speed of the Earth is 3,074.6 m/s). When we plug in the values for G, M_E, and v, we get a r value of 42,164 km. When we subtract the radius of the Earth at the equator (6,378 km), we get 35,786 km, which is the geostationary altitude.

Once we reach lower altitudes, orbit speeds start to increase,[33] meaning that satellites are not geostationary anymore. They spin faster than the Earth. Between 2,000 and 36,000 kilometers (1,200 and 22,200 miles), we are in the medium Earth orbit (MEO). MEO satellites have especially been used for geolocation—we do not want to communicate with only one satellite anymore but several to get an exact location. The famous Global Positioning System (GPS) operates thirty-one satellites that can tell you virtually everywhere around the planet where you are located. They orbit at an altitude of around 20,000 kilometers (12,400 miles). Since the original GPS that is operated by the United States, many countries have developed their own satellite positioning systems. At this altitude, two-way communication speeds are a bit faster, though not fast enough to browse the Web seamlessly. For this task, we need to go lower.

We now enter the world of low Earth orbit (LEO) satellites. LEO altitudes range from 160 to 2,000 kilometers (99–1,200 miles). At this altitude, we are now much closer to the Earth. The most famous LEO satellite is the International Space Station that orbits the Earth at an altitude of 400 kilometers (250 miles). From a GEO satellite, a signal takes around a quarter of a second (or roughly 250 milliseconds) to reach the Earth. With a LEO satellite, it takes only around one two-hundredth of a second (or roughly five milliseconds), which is fifty times faster.[34] Now we can browse the Web with *low latency* (i.e., little delay since the signals can go back and forth quickly). In urban areas, fiber will always be faster and preferable, but in locations like rural areas,[35] ships, and airplanes that do not have easy access to a fiber-optic cable, LEO satellites are ideal. A little like in the story of Oaxaca, LEO satellites enable internet service providers to adopt a new business model, one that does not discriminate against people who live in remote areas.

The caveat is that the area a single LEO satellite can cover is dramatically reduced. From tens of satellites at MEO and GEO altitudes, we now need thousands of satellites to cover the planet. A term used to describe these types of systems is "satellite internet constellation."

Many of the values reported here come from the May 1, 2023, Union of Concerned Scientists (UCS) Satellite Database. Out of 7,560 satellites reported, the database lists 590 GEO satellites, 143 MEO satellites, 6,768 LEO satellites, and 59 elliptical satellites.[36] Yet several private companies had already submitted applications to the United Nations for tens of thousands of new LEO satellites, which is not without concerns. LEO satellites reflect light, significantly affecting space

---

33. Rearranging the equation in the previous footnote, we get $v = \sqrt{\dfrac{G \cdot M_E}{r}}$. We can see that as r decreases, v increases.

34. These values are approximate. Signal transmission times vary significantly depending on altitude.

35. While prices remain expensive as of this writing, LEO satellites offer a new business model that may fit people like the communities in Oaxaca featured at the end of the previous chapter.

36. Not discussed here. They are mostly used for Earth observation and space science.

observation. Moreover, more satellites mean a higher probability of collision and greater chance of debris cluttering the sky.[37] We must also be aware that sending thousands of satellites into space generates an enormous amount of greenhouse gases at a time when we need to limit our emissions as much as possible.

From the 1950s, when satellites belonged mostly to the realm of science fiction, they have become an integral part of the silent and seamless infrastructure that enables our lives.

---

37. I just hope a space dumpster truck will never have to be invented.

Part Five

INFRASTRUCTURE

CHAPTER SIXTEEN

Infrastructure

We Have Come Full Circle in Singapore

MAJULAH SINGAPURA! ALTHOUGH THE ENTIRE CITY-STATE[1] OF SINGAPORE may only spread over an area of 725 square kilometers[2] and have a population of slightly fewer than six million people, it is arguably the city with the most developed infrastructure in the entire world.[3] Thanks to its achievement, and despite its small population and size, virtually everyone in the world has heard of Singapore (even if few can locate it on a map). Singapore leaped from being a relatively poor place in the early 1960s to being one of the richest nations by the early 2000s. What Singapore has achieved in forty years is incredible. I think that the title of the national anthem, "Majulah Singapura" (which translates from the Malay language as "Onward Singapore"), fits Singapore perfectly.

I had the chance to live one year in Singapore, from August 2011 to July 2012. I was stunned by just about everything in the city-state. It was

---

1. Technically, Singapore is both a city and a country. It is the size of a city, but it is independent like a country. A common term for these types of places is city-state, but Singapore is unique. Other city-states include Vatican City and Monaco, but they do not have customs and immigration borders with their neighboring countries like Singapore does. Other places like Dubai, Abu Dhabi, Hong Kong, and Macau are viewed as city-states as well, but they are not independent like Singapore is.

2. As a comparison, the small European country of Luxembourg is nearly 3.5 times larger, with 2,590 square kilometers. Hong Kong is 3.8 times larger, with 2,755 square kilometers. The small U.S. state of Delaware is nearly seven times larger, with 5,130 square kilometers.

3. Several smart cities have been built, but they tend to be located around existing urban areas. As a livable city with a rich history, in my mind, Singapore has the most developed infrastructure as of this writing.

always clean and safe. Despite their differences, the three main ethnicities—Chinese, Malay, and Indian—cohabited peacefully, and I enjoyed celebrating Chinese, Muslim, Hindu, Buddhist, and Christian holidays. Virtually every family was relatively well off and owned a property thanks to the Housing Development Board (HDB), a government agency that plays a big part in the city-state. Despite its small size, Singapore featured tons of things to do. The local cuisine was unique and fantastic. In fact, to this day, one of my all-time favorite meals comes from Singapore. It is a tasty peppery broth with pork ribs called *bak kut teh*[4] that is served with rice and *youtiao* (fried dough sticks); just thinking about it makes me salivate. And, of course, Singapore's infrastructure operated nearly perfectly. Life was as easy as it gets.

After being a British colony for one and a half centuries, and then part of Malaysia for two years, Singapore was thrown into independence in 1965. It had to fight to survive, and it fought well. With Lee Kuan Yew at the helm as the founding prime minister, Singapore became a major maritime transport hub in the world, an industrial center, a financial center, and a technology hub. Because of its small size and lack of resources, Singapore had to be smart and creative, and it managed to turn its constraints and limitations into assets. A little like the story of Bogotá's transit system, Singapore did not seek to apply best practices; instead, it created its own solutions. While the prospect of a better job and not so hot and humid weather led me away from Singapore, it will forever be a special place for me.

In this chapter, we come full circle and apply all the infrastructure systems that we have covered to one single city. We will learn about the water, wastewater, transport, electricity, gas, solid waste, and telecommunication systems of Singapore.

In terms of water collection, we learned that the most important force to master is gravity (to enable the transport of water over long distances

---

4. It literally translates to "pork bone tea." There are two kinds of broth, peppery and herbal; I prefer the peppery kind, and I have a favorite joint in Singapore. *Bak kut teh* is always the first meal I have when I visit Singapore, and I cook it at home regularly.

without pumps). Singapore is small and does not have access to large freshwater sources. Like Hong Kong, which receives water from China, Singapore imports fresh water from Malaysia through a pipeline. Fresh water is stored in several reservoirs located across the city-state. Historical reservoirs are located at the center of the island, where elevation is highest, to make use of gravity for distribution. But the most famous reservoir is located at a low elevation, by the sea,[5] at the estuary of the Singapore River in downtown Singapore. It is the Marina Reservoir, which opened in 2008. It is surrounded by reclaimed land that features the iconic Marina Bay Sands. The reservoir was given a horseshoe shape to store water, and a barrage stops the flow of fresh water from the estuary to the sea. Several such reservoirs were built in Singapore.

As water is scarce, restrictions and economic incentives have been implemented to control water consumption, which stood at 149 liters per day per person in 2023—similar to that of Israel. Even then, collecting rainwater and importing fresh water is not sufficient. Like Israel, Singapore turned to desalination, using reverse osmosis to turn seawater into pure water by pushing it through a membrane with pores that are so small that even diluted minerals like salt cannot get through. Water is then re-mineralized, and chlorine is added as a final disinfectant before distribution.

To treat fresh water, Singapore uses the conventional sequence of screening, coagulation, flocculation, sedimentation, filtration, and disinfection, like the processes described in the chapter on water treatment in Tel Aviv. Screening removes particles larger than one millimeter; chemicals are then added to favor coagulation (the neutralization of the electronic charge of suspended particles) and flocculation (the formation of larger clumps); the clumps subsequently settle at the bottom of the chamber (sedimentation) and are removed; the clear water is then filtered[6] to

---

5. As we have seen, Hong Kong also has several reservoirs located by the sea.

6. In Singapore, sand is used as well as membranes similar to reverse osmosis, but the pores are larger so that less pressure (i.e., energy) is needed to let minerals pass through.

remove fine particles, and chlorine is added here as well as a final disinfectant before distribution.

Fresh water is collected from precipitation and imported from Malaysia, and seawater is desalinated, but Singapore went one step further. It also recycles some of its wastewater.[7] In Singapore, about 45 percent of the water is consumed for domestic purposes (i.e., households) and 55 percent for nondomestic purposes, including many industries that require water, but the water does not need to be potable. Branded as NEWater, in 2019, around 25 percent[8] of the total water needs were supplied with recycled water, mostly to industries that do not require potable water. Some water is also reserved for indirect potable use by being added to the city-state's reservoirs during the dry season to go through the conventional water treatment process later.

The treated water is distributed through a network of 5,500 kilometers of pipes at a pressure around 25 meters, like that of many cities. While Singapore may very well have one of the newest and best-maintained water distribution systems in the world, leakage remains inevitable. In Singapore, leakage was estimated to account for 8 percent of the water used. This number is significant. It really goes to show how leakage is unavoidable (including in natural gas systems).

For wastewater treatment, Singapore adopted a conventional system as well. In the chapter on sanitary sewers in Paris, we learned about four steps: pretreatment, primary processes, secondary processes, and tertiary processes. Pretreatment processes make use of screens and a grit chamber to rid water of particles roughly larger than a grain of sand and of some greases. Primary treatment processes push the goal of pretreatment further by using chemicals and allowing for fine particles to settle (flocculation) and for greases to float up to the surface in large tanks called primary clarifiers. Secondary treatment processes leverage biofilters to handle

---

7. As we have seen, in Paris, the treated non-potable wastewater is used for irrigation, to clean the streets, and to flush the sewers (for maintenance).

8. You can read more about water circularity and Singapore in an article I coauthored: M. Arora, L. W. Yeow, L. Cheah, and S. Derrible, "Assessing Water Circularity in Cities: Methodological Framework with a Case Study," *Resources, Conservation and Recycling* 178 (2022): 106042.

dissolved particles by aerating the wastewater and using microorganisms that consume the dissolved particles for their personal growth. No conventional tertiary processes are applied in Singapore. After some measurements are made to ensure that the wastewater respects strict environmental standards, the treated wastewater is discharged into the sea. But not all of it. Some of it is sent for further treatment, using advanced membrane and reverse osmosis to make it high quality for industrial use. A new life for NEWater begins as described before.

At this point you might be wondering what type of wastewater system Singapore has. In Paris, we discussed how separate sewer systems are better than combined sewer systems because heavy rain events can overwhelm combined sewer systems, leading to sewer backups and the disposal of raw sewage in water bodies. It may not come as a surprise that Singapore has a separate system. Therefore, only the sanitary wastewater is treated (termed "used water" in Singapore). To transport sanitary wastewater to one of the wastewater treatment plants, Singapore also built a series of large deep tunnels.

Having a separate sewer system does not mean that stormwater management is not important. While it prevents raw sanitary sewers from being disposed of in natural water bodies, infrastructure must be built to handle stormwater. In fact, effective stormwater management is imperative in Singapore, as it must deal with torrential rains that always impressed me by how scary they were—the loud thunderstorms and furious lightning would not be out of place in a scene from a Hollywood movie. Historically, Singapore had built an effective gray infrastructure system by installing drains underground and having small ditches on the sides of many roads to carry stormwater to larger open (concrete) canals and rivers to rid the city of stormwater as quickly as possible. From our time in Copenhagen, you may remember that these methods have limits, especially in growing cities where new roads and buildings are constructed, making larger surface areas impermeable. It did not take long for Singapore to embrace green infrastructure principles and implement low-impact development strategies, including permeable pavers, green roofs, rain gardens, and bioswales, in addition to building detention and retention basins. Bishan-Ang Mo Kio Park offers a fitting example. The park was completely redone and opened in 2012, and it features many effective stormwater management strategies. Using gravity, stormwater is managed and transported, but it is not discharged into the sea. Instead, it is kept in reservoirs, like the Marina Reservoir, to be treated and distributed later as drinking water.

The entire water/wastewater system in Singapore is remarkable. Through creativity and engineering, Singapore managed to build some of the best water/wastewater systems in the world.

If you are impressed by Singapore's water/wastewater system, just wait to learn about its transport system, which is, in my mind, currently the best in the world. I consider Singapore's transport system the best because it emphasizes both mobility and accessibility. It is both easy to travel around Singapore and *reasonably* inexpensive (notice the emphasis on *reasonably* because there is a catch).

First remember the concepts of speed, density, and flow that we discussed in the chapter on traffic and roads in Los Angeles, and while individuals try to maximize their speed (to minimize their travel time), traffic engineers focus on maximizing flow. Second, we remember the inverted U-shape relationship that tells us that flow is maximized when density is around 20–25 percent of the jam density. To enhance mobility, density must therefore be controlled. High density (i.e., congestion) is a sign of a high demand. We also remember that demand is driven by two elements: number of users and trip distance. To manage demand, since 1990, Singapore has controlled the number of users by making it prohibitively expensive to own a private vehicle. Before owning a vehicle, individuals must purchase a certificate of entitlement that is attributed through a bidding process. In 2023, the cost of a certificate of entitlement was around US$80,000,[9] and these certificates are valid for only ten years (after which

---

9. People submit bids to acquire a certificate of entitlement, and therefore prices vary all the time (and can increase rapidly). Prices also vary depending on the category of vehicle class (i.e., small versus large vehicles, commercial vehicles, and motorcycles). The example featured here corresponds to a category A vehicle (the smallest type). Search online for "Singapore Certificate of Entitlement (COE)" to find the latest prices on the website of the Land Transport Authority.

point a new one needs to be purchased). But the story does not end here. In addition to the certificate of entitlement, individuals must pay a small registration fee, an additional registration fee (that costs at least 100 percent[10] of the value of the vehicle), an excise duty, and the usual sales tax; vehicle emission rebates are available for low-emission vehicles. One source mentioned that a US$13,400 vehicle cost around US$114,000 in Singapore in 2023.[11] What is more, these prices apply to the purchase of the vehicle only. Gas and parking prices are also high. An annual road tax must be paid. Maintenance costs can quickly escalate. And then you have the Electronic Road Pricing (ERP) system. ERP is a toll collection system that consists of gantries (i.e., gates). Each time a vehicle passes through a gantry, it must pay a fee that changes during the day to control congestion. Wikipedia[12] reported ninety-three gantries in Singapore as of July 2024.

Owning and using a car is therefore, again, prohibitively expensive. Yet I mentioned earlier that Singapore emphasizes mobility. This is because Singapore has a stellar public transport system. From our experience in London, we remember that public transport is preferable to private transport because transit riders take up less physical space than people driving private vehicles. In 2024, Singapore had 6 metro lines (more than 140 stations), 3 light rail transit lines (more than 40 stations), and more than 370 bus lines (more than 4,500 stops). At least two more metro lines are expected to be built. The metro system is called the MRT for Mass Rapid Transit. The Singapore transit system is operated mainly by two private companies: the Singapore Mass Rapid Transit company and the Singapore Bus Company—the names are misleading since both companies run both metro and bus lines. In terms of ridership, in 2023,

---

10. Say the open market value of the vehicle is $10,000; in this case, the additional registration fee costs another $10,000. In 2023, the rates were roughly as follows in US dollars: 100 percent for vehicles less than $15,000, 140 percent for vehicles between $15,000 and $37,500, 180 percent for vehicles between $37,500 and $60,000, and 220 percent for vehicles more than $60,000.

11. And the costs keep increasing (especially the certificate of entitlement). I first wrote this text in summer 2020; at that time, a US$12,000 compact car cost roughly US$52,000.

12. https://en.wikipedia.org/wiki/Electronic_Road_Pricing (accessed August 25, 2024).

the metro system carried over three million riders per day, the bus system carried over 3.5 million people per day, and the LRT carried over two hundred thousand riders per day.[13] These numbers are impressive for a population of about six million people. The Deloitte City Mobility Index 2020 states that the mode share distribution in Singapore is as follows: 33 percent private vehicles, 53 percent transit, 12 percent walk, and 2 percent bike.[14]

The Singapore transit system has another important mode that is not always considered transit but is still essential: taxis. I seldom used taxis before moving to Singapore, but it became a habit when I lived there, in part because they were surprisingly affordable. And I was not alone: taxis served over six hundred thousand daily riders in early 2024. While I would walk or take transit for maybe 95 percent of my trips, I often took a taxi when transit was not an option or later in the evenings. And I would also pay with my EZ-Link card. In Singapore, the EZ-Link card can be used to pay for transit, taxis, gasoline, parking, and many other things, including ERP tolls. Overall, during my stay, I never wished I had a car, and I never felt limited in terms of traveling—I was as mobile as I desired.

Singapore realized a long time ago that traffic congestion has a significant and negative impact on the economy, livability, and public health (in terms of air quality) of its residents. Controlling congestion by restricting car ownership while offering an extensive, convenient, and affordable transit system provides the ingredients for a thriving economy and enhanced livability. But we are not done with transport yet. To emphasize accessibility, Singapore also carefully monitors how land is used. In Shanghai, we saw that land use has a significant impact on demand for transport. Mixed land-use policies (i.e., a mix of residential and commercial buildings) are generally desirable so that essential amenities (e.g., shops,

---

13. You can find the latest numbers on the Land Transport Authority (LTA) website; search online for "Singapore LTA statistics."

14. The hot and humid Singapore climate does not make it conducive to biking.

grocery stores, and restaurants) are within a short walk. In Singapore, you are never far away from a *Hawker center* (i.e., an outdoor food court), a place to enjoy delicious chili crab, chicken rice, *nasi lemak, murtabak, roti prata, kaya* toast, *kopi*, or my favorite, *bak kut teh*. The Singapore government has an amazing public housing program administered by the Housing Development Board (HDB), which builds large residential complexes with commercial space so that most amenities are available within a short walk. The HDB apartments are subsidized, and, as a result, most Singapore families own their property.[15] Thanks to these land-use policies, Singapore manages to have a relatively high walk-mode share distribution despite the hot and humid climate, and much credit goes to the Land Transport Authority (LTA) and to the Urban Redevelopment Authority (URA).[16]

If I have one criticism about Singapore's transport system, it is its lack of shared space. All modes are too carefully segregated and the use of elevated walkways to cross busy streets is prohibitive to walking, but as complete streets and shared spaces become more popular, I look forward to seeing how it will impact the city-state.

So far, I have offered glowing comments about Singapore's infrastructure, but my tone is about to change with electricity. Let us first recall the concepts of electricity we learned in Cape Town.

Because electricity is not naturally available in large quantities on Earth, another form of energy must be harvested and converted into electricity. The main way to generate electricity is through an electric motor that converts mechanical movement into electricity. Therefore, raw sources of mechanical energy such as rivers, the wind, and tides are preferred since

---

15. But only Singapore citizens and permanent residents can buy them (and get low government-sponsored interest rates). They also have to earn less than a certain income to be eligible. Moreover, as the demand is high, single people cannot buy an HDB apartment before turning thirty-five years old. Race quotas are also enforced so that every residential complex has a mix of Chinese, Malay, and Indian people. Finally, technically, the apartment is not "bought" but owned for ninety-nine years only. While these rules may sound odd at first, the system works extremely well.

16. If you visit Singapore, I recommend visiting the URA building, which has a great, free exhibition called the Singapore City Gallery, featuring a large model reproduction of the city-state.

energy is converted only once (i.e., higher efficiency). In the absence of raw mechanical energy, another form of energy needs to be tapped first. This is where fossil fuels come into play. Fossil fuels possess dense quantities of chemical energy that, when burned, convert into thermal energy (i.e., heat) that can be used to rotate a turbine (i.e., mechanical energy) either directly (e.g., natural gas) or by using steam (e.g., coal). Nuclear energy applies here as well since nuclear fission produces heat. The main alternative to generate electricity is through solar photovoltaic—photons excite semiconductive material like silicon, energizing it to release electrons (i.e., electricity).

We also compared methods to generate electricity through three criteria: efficiency, operating characteristics, and sustainability. Efficiency is all about the number of times energy is converted (fewer conversions means fewer losses), and therefore raw sources of mechanical energy such as hydroelectricity are preferred here as well. Operating characteristics focused on whether the energy source was flexible (i.e., controllable) or intermittent (i.e., uncontrollable). This is where solar and wind lose ground to other methods since the weather is not controllable, although the power grid can be set up so that flexible power plants adjust their output in relation to the maximum capacity of solar and wind farms at any given time. Plus, we are getting better at storing electricity (e.g., using pumped storage and batteries) that can serve as virtual power plants during peak demand, and we are getting better at managing the demand by asking customers to consume less during peak demand (e.g., by adjusting their thermostats). Finally, sustainability is related to how much the consumption of an energy source can be sustained over time. Here, renewable sources of energy like hydro, solar, and wind are preferred. In contrast, nuclear energy is unsustainable since nuclear fuel (e.g., uranium) is available in finite quantities—that is, once we run out, we cannot produce nuclear fuel. But the prize for unsustainability goes to fossil fuels since they are unsustainable on two counts. Beyond existing in finite quantities on Earth like nuclear fuel, fossil fuels, when burned, generate greenhouse gas emissions, and humanity has burned so much fossil fuel that it has changed the climate. The use of fossil fuels to generate electricity must therefore be phased out. We also have an internal ranking of the most unsustainable fossil fuels. Coal ranks worst, followed by oil (75 percent as bad as coal), and then natural gas (50 percent as bad as coal). The clear winner in terms of electricity generation is hydroelectricity, and the biggest loser is coal. As sources of renewable energy (thus sustainable), solar and wind are preferred.

Based on this information and on the fact that Singapore is small, can you guess how electricity is produced in the city-state? This is where it gets disappointing. If I tell you that Singapore uses natural gas, it might not come as a surprise, but that's how Singapore produces 95 percent of its electricity. Disappointingly, solar energy is barely used despite the fact that Singapore receives abundant quantities of sun. The fact that Singapore uses natural gas to generate electricity at such a high rate makes it a high per capita emitter of greenhouse gases. One solution is to invest in renewable energy and in energy storage (to create a virtual power plant that can provide electricity during peak demand or when there is less or no sun, for example), but land is scarce in Singapore. An additional possible solution is to build a regional power grid across Southeast Asia so that water-rich countries like Laos and Vietnam with high hydroelectricity potential transmit cleaner electricity across the entire region. In 2020, there were even talks to build a solar farm in Australia and transmit the electricity to Singapore through close to four thousand kilometers of undersea cables.

After being generated, electricity is stepped up between 230,000 and 400,000 volts before being transmitted. Because land is so scarce in Singapore, virtually all infrastructure, including electricity transmission cables, are buried underground, sixty to eighty meters deep so as not to interfere with other underground infrastructure. This transmission network spans forty kilometers, and the tunnels have diameters between six and eleven meters. These numbers are impressive.[17] Nearly thirty thousand kilometers of distribution cables are also buried, but at much shallower depths, as is common, and voltages are stepped down across the network until they reach 230 volts at individual outlets.

After electricity, we covered natural gas. While Singapore does not have a natural gas distribution system, it receives a lot of natural gas for electricity generation. We can recall from our visit to Buenos Aires that, after gas is extracted, it undergoes four stages. First, liquids are removed. Second, all water vapor is removed to dehydrate the gas. Third, heavier hydrocarbons with higher market values are separated so that natural gas consists mainly of methane. Fourth, final impurities are removed, including carbon dioxide and sulfur. The separation of these constituents during the four stages is done mostly through adsorption—that is, a solid material that reacts with a desired constituent is used to trap it. Dry gas is then sent over pipelines or liquified and shipped on gas tankers. By playing with pressures in pipelines, linepacking can be increased or decreased depending on the demand at the destination. In Singapore, instead of being stored in natural underground storage space, the imported liquefied natural gas is contained in enormous, cooled tanks before the liquid is turned into a gas again and burned. Yet Singapore does use its underground to store liquid petroleum products (i.e., to store oil)—the Jurong Rock Caverns were built for that purpose. The argument is always the same: drill the underground to save space on the surface. The Public Utilities Board—Singapore's national water agency, which manages the city-state's water and wastewater infrastructure described above—is also looking into building deep underground freshwater reservoirs.

How does Singapore handle its solid waste? Recall the concepts learned in the chapter on solid waste management in Tokyo. First comes the integrated solid waste management hierarchy: source reduction and

---

17. And I would not recommend it for other places since building it must have required a tremendous amount of energy, and it must have been costly, but it demonstrates the value of the land in Singapore.

reuse, recycling and composting, energy recovery, and treatment and landfilling. The primary goal is not to generate solid waste in the first place. In this respect, like Tokyo, Singapore is doing well with twenty-seven kilograms of solid waste per person per month in 2020.[18] In terms of recycling, Singapore is not doing well. Domestic recycling rates were at 12 percent in 2023, although, like Tokyo, the post-processing of the recyclable waste is partly at fault.[19] The solid waste that is not recycled is sent to one of the waste-to-energy recovery facilities (i.e., incinerators) for burning to produce electricity (contributing to the 5 percent that does not come from natural gas). The resulting ash and other nonburnable solid waste with no value (including the sludge from the water and wastewater treatment plants) are sent to the Tuas marine transfer station to be sent to the Semakau landfill. The particularity of the Semakau landfill is that it is an island located eight kilometers south of the mainland. Like in Tokyo, nonrecoverable waste is used for land reclamation. The construction of the Semakau landfill started in 1995. A seven-kilometer perimeter bund (embankment) was first constructed in the sea, encircling two small islands. The bund was lined with an impermeable membrane and a layer of marine clay so that any potential leachate does not contaminate the sea. Sand bunds were then added to the area to divide it into cells to be filled up one at a time. The Semakau landfill began operation in 1999, and part of the island is open to the public as a park. The Singapore National Environment Agency has published relevant information and made some

---

18. In my book *Urban Engineering for Sustainability* (Cambridge, MA: MIT Press, 2019), I quoted a value of 115 kilograms that I had obtained from a World Bank report and that appeared much too high to me. As expected, the number was off, but I am not sure whether it is because total solid waste data were included as opposed to municipal solid waste only (i.e., not including construction and demolition debris and industrial waste) or whether it is because the World Bank had estimated municipal solid waste through a model that inflated Singapore's number because the city-state enjoys a high gross domestic product per capita. This number cited here comes from the same source used in the solid waste chapter, which is an updated version of the earlier report that misestimated Singapore's value. The report is S. Kaza, S. Shrikanth, and S. Chaudhary, *More Growth, Less Garbage*, Urban Development Series (Washington, DC: World Bank, 2021).

19. That is, Singaporeans sort much of their waste, but not all of it gets recycled. The value cited here comes from the newspaper article: A. Yu, "Singapore's Domestic Recycling Rate Stalls at 12%, the Lowest in over a Decade," *Straits Times*, June 19, 2024, available at https://www.straitstimes.com/singapore/domestic-recycling-rate-stalls-at-12-despite-decline-in-household-waste (accessed August 25, 2024).

helpful videos to explain the construction and operation of the Semakau landfill.

We are now down to our last infrastructure system: telecommunication. In a highly connected world, any country or city that wants to be competitive in the global market must have a top-notch telecommunication system. Singapore understood this point a long time ago and has purposefully become a telecommunication and technology hub. We remember that picking up an analog phone enabled an electrical current, initially to one single phone only, then to an operator who would use a physical cable to establish a direct line between a caller and a receiver, and later to the local service switching point (switch) that would query the parallel computer system (Signaling System 7; SS7) whether a caller's number was blocked by a receiver and, if not, how to route the call. The intensity of the electrical current would go up and down as the frequency of the people's voices went up and down. To fit multiple calls in one single cable, frequency-division multiplexing (FDM) was developed. Voice calls were contained in four-kilohertz (kHz) bands. The television system worked similarly, albeit by using six-megahertz (MHz) frequency bands to be able to transmit both the audio and the video. This need for higher frequency bands also explains why television channels are transmitted through (larger) coaxial cables and telephone calls are transmitted through (smaller) twisted pairs of wires.[20]

In a wireless world, both telephone calls and television channels can also be transmitted over the air by playing with the frequency of electromagnetic waves—look for broadcasting antennas on skyscrapers or towers and for white rectangle boxes on tall buildings in your neighborhood. Over time, the air has become a big mess of electromagnetic waves. In Singapore, the Infocomm Media Development Authority (IMDA) regulates the radio spectrum. Figure 16.1 shows Singapore's spectrum

---

20. If you check your landline or DSL jack, you will see four wires as opposed to two so that homes can handle two phone lines if they like.

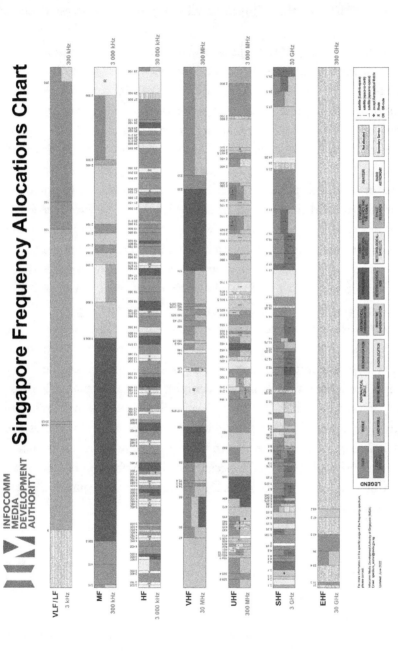

Figure 16.1. *Singapore Frequency Allocations Map* (*June 2022*).
Infocomm Media Development Authority (IMDA) of Singapore.

allocation chart (i.e., the confusing mall map). It can be accessed freely online.[21]

Everything changed with the digital revolution, as we learned in San Francisco. The big innovation was the conversion of a complex frequency signal into a series of binary values—on or off—which are extremely easy to compute and transmit and are less prone to errors.[22] In particular, as the fastest element in the universe, light (and the absence of light) can be used to transmit a binary signal. The unit of measure (on or off) is called a *bit*, but a bit by itself cannot do much. To convey complex information, a series of bits needs to be put together to mean something. Internationally, it was decided that a series of eight bits, called an *octet* or a *byte*,[23] would be used. A suite of protocols was needed to turn the transmission of octets into meaningful information. The most common representation of digital telecommunication in a computing system is the Open Systems Interconnection (OSI) reference model that has seven layers: physical, data link, network, transport, session, presentation, and application. We will not go through the seven layers again, but we can recall that long strings of octets that convey complex information are divided into small packets that use internet protocol (IP) addresses to make their way through the impressively large and complex physical layer that is composed of routers, public Wi-Fi devices, cell towers, internet exchange points, fiber-optic cables, satellites, and so forth. Speaking of satellites, the May 1, 2023, UCS database reports that Singapore operated or owned nine satellites (all LEO), used especially for Earth observation and technology development. It also shares four more with other countries, and it operates several for other countries.

---

21. The figure shows the June 2022 version of the spectrum allocation chart (latest available at the time), collected from https://www.imda.gov.sg/-/media/imda/files/regulation-licensing-and -consultations/frameworks-and-policies/spectrum-management-and-coordination/spectrumchart .pdf (accessed May 29, 2024).

22. Because the signal must be repeated every few kilometers and a simpler signal is easier to repeat.

23. But the length of a byte has evolved over time and has been used for a combination anywhere between one and forty-eight bits, which is why we favored the term "octet."

Singapore has rapidly become one of the most connected nations in the world. By 1994, it was already one of the first countries to have a fully digital telephone network. By 2020, there were close to nine million mobile phones in the city-state (about one and half per person), and virtually everyone had access to broadband internet. Singapore is so invested in telecommunication that the IMDA set up a massive, free Wi-Fi system around the city-state, mainly in commercial areas, through its program Wireless@SG. To stay ahead and offer cloud services, Singapore also initially built many data centers but then stopped after realizing how much electricity data centers require (and we saw above that Singapore can do better with its electricity generation).

Overall, Singapore's telecommunication system can be praised, along with its water and wastewater system, its fabulous transport system, and some of its solid waste management practices. Singapore is lacking in how it generates electricity and in its poor recycling rates, but I hope to see changes in the future.

We have covered a lot over these pages. I hope to have inspired you to look at infrastructure in a different way and even to look for it around you. That is what I did many years ago when I returned to Saint Pierre, my hometown. Geographically and culturally, it is diametrically opposed to Singapore, but in terms of infrastructure, it is not that different, as we will see in the story below.

We have almost reached the end of our journey. In the concluding remarks, we will remind ourselves why we build infrastructure in the first place and discuss the future of infrastructure.

THE INFRASTRUCTURE OF SAINT PIERRE AND MIQUELON

I was born and raised in Saint Pierre, one of two municipalities in the archipelago of Saint Pierre and Miquelon, located just south of Newfoundland. Saint Pierre is the smaller island, but it is where most people (roughly fifty-five hundred) live. Miquelon is home to closer to five hundred people, giving us a grand total

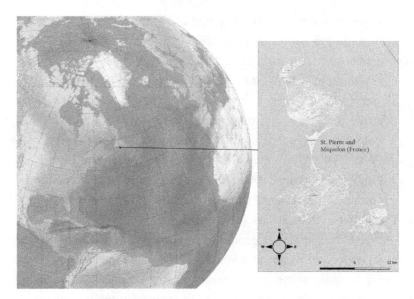

Figure 16.2. *Map of Saint Pierre and Miquelon.*

population of around six thousand. Having a small population is not the most interesting feature of the archipelago, however. The archipelago is a France-owned territory, meaning that I am a French citizen. Saint Pierre and Miquelon is technically part of Europe. Yet geographically it is located on another continent.

The archipelago is nineteen kilometers (twelve miles) off the southern coast of Newfoundland (see figure 16.2[24]). The larger island to the north is Miquelon. It is linked to the island of Langlade[25] to the south by a sandy isthmus that was formed over time in part thanks to shipwrecks.[26] Southeast of Langlade, you have the island of Saint Pierre, where I was born and raised until I was eighteen years old (like most, I had to leave to go to university). Although I am clearly biased, I find the history of the archipelago riveting. It includes native populations, wars between France and the United Kingdom, American Prohibition, World War II, and a fishing moratorium that transferred the economy of the archipelago to infrastructure development.

---

24. This map was produced by Juan Acosta Sequeda, who was my PhD student at the time. It combines basemaps from Natural Earth and Open Street Map.

25. Where many people from Saint Pierre have cabins and vacation homes to spend their weekends and holidays.

26. Captains thought they could cross the small channel, but they could not. The ships offered the structural elements needed to hold the sand.

While Saint Pierre and Miquelon is in North America, the culture is surprisingly very French. I think infrastructure has a lot to do with this situation since it follows French and European regulations. Plus, French and European funding sources are sought to finance infrastructure. Examining Sain Pierre through the lens of its infrastructure, we can learn a lot about its people, history, and future.

Starting with water, it may not come as a surprise to learn that the people of Saint Pierre live closer to the shore, in (almost by definition) lower-lying areas. Yet Saint Pierre is hilly. Large ponds at higher elevations were dammed and turned into freshwater reservoirs. The big advantage is that pumps are not needed since gravity can do all the work to distribute water. Saint Pierre also has a small water treatment plant located by the main freshwater reservoir that was built in the 2000s.

In terms of wastewater, Saint Pierre followed the global trend. The sewer system in the older part of the city is combined, while the newer parts of the city have separate sewer systems. Historically, all wastewater was dumped into the ocean with no treatment. In the 2000s, sewer conduits were installed by the shore (at the lowest elevation) to collect all the wastewater before it gets dumped into the ocean and send it to a wastewater treatment plant. The system never worked,[27] however. As of this writing, wastewater is still dumped into the ocean with no treatment. Let us hope the situation evolves by the time you read this story.

Next comes transport. As it is small, Saint Pierre has no traffic signals/lights and no public transport system. That said, there are several taxi operators, and there is a school bus system to transport kindergarten and elementary school children. To manage traffic, in the absence of a traffic sign, everyone obeys the priority to the right rule: at every intersection, drivers slow down, look to their right, and, if there are no vehicles, cross the intersection; otherwise, they let all vehicles to their right go first. Following French law, people must also be eighteen years old or older to get a driver's license.

When it comes to electricity, Saint Pierre is powered by six large diesel generators that can provide a total capacity of twenty megawatts. Peak power demand is closer to ten megawatts, but doubling the capacity is a resilience measure. While fossil fuels should be avoided to generate electricity, there are currently no alternatives being used in the islands. A flexible electricity generation method is needed, and there is no capacity for a hydroelectric power plant or a small nuclear power plant. In the future, when it is easier to store electricity and create virtual power plants, harvesting the energy from the wind and the tides[28] should offer

---

27. The system had flaws. The pumps never functioned properly. The upgrade would be expensive and would need to be paid either by the City of Saint Pierre or by the regional government of Saint Pierre and Miquelon, but no one wants to pay for it. This story is typical. Sewer infrastructure is neither urban nor regional. That's why sewer district authorities are created.

28. We have not discussed electricity generation from the tides much in this book. That is mainly because it is not easy, especially as salt is corrosive and damages electrical equipment.

viable options. I should also point out that toxic emissions from the diesel motors are captured instead of being released in the atmosphere. Moreover, a district heating system was built to recuperate the heat from the diesel motors and use it to heat the administrative buildings around the island. The electricity is distributed through 13-kilovolt power lines (a large transmission system is not needed). Naturally, as Saint Pierre is a French territory, all buildings are supplied with 230 volts.

Neither Saint Pierre nor Miquelon has a natural gas system. We remember that liquifying natural gas and then storing it is challenging. It does not make sense for a small population. For heating, many buildings are equipped with electric radiators or oil furnaces[29] (while it costs more, liquid oil is easier to ship and handle than natural gas).

For the longest time, solid waste was stored and sometimes burned in an open-air dump on one side of the island. A small revolution happened in the 2010s, however. New infrastructure was built, and now residents are asked to sort their solid waste into six categories: glass, plastics, cardboard, food waste and bio waste (e.g., untreated wood), electronics, and everything else. Glass is crushed and used to make pavement. Plastics and cardboard are compacted and sent to Canada for recycling. All food and bio waste are composted locally. All electronics (which run on 230 volts) are sent to France for recycling. Finally, what is left is burned or left in the open air, but there is talk about building an energy recovery facility.

The last infrastructure we learned about is telecommunications. Saint Pierre has had a traditional telephone system for a long time. Internet used to be provided via an antenna located in the south of Newfoundland. However, a submarine fiber-optic cable was installed in the 2010s and now physically links Saint Pierre to Miquelon and to Newfoundland (forming a triangle). As a last point, Saint Pierre also hosts a ground station for the European global navigation satellite system called Galileo—Saint Pierre and Miquelon is the only European territory in North America.

In the end, despite the small population, the infrastructure of Saint Pierre is fairly intricate and well thought out. As I frequently return to Saint Pierre, I will keep following how the infrastructure of the entire archipelago evolves, and I cannot wait to see how it becomes more sustainable, resilient, and livable.

---

29. Technically, similar to what we discussed about gas furnaces in the Cape Town chapter, oil furnaces generate fewer greenhouse gas emissions in Saint Pierre since electricity is generated by burning diesel.

Concluding Remarks and
the Future of Infrastructure

In their song "Beyond,"[1] the band Daft Punk has the following verse: "The perfect song is framed with silence." I like this verse a lot. It gives silence—that is, the absence of music—equal importance to music itself. I would like to tweak this idea and apply it to infrastructure and society: the perfect life and the perfect society are framed with infrastructure.

As discussed at the outset of this book, we tend to forget about infrastructure and take it for granted. Infrastructure tends to be seamless and silent, out of sight and out of mind. We seek instantaneous ease without really understanding it. We must also remember that all infrastructure is made by humans and for humans. That is why infrastructure exists. I want to emphasize the "for humans" part as infrastructure should always be built to help us connect and provide for us. In Antoine de Saint-Exupéry's *The Little Prince*, when the fox meets the little prince, he tells him in his own way that the meaning of life is "to establish ties" between one another. Is it not spot on? As humans, we seek to get together, establish relationships, build something together, and infrastructure enables us to do that. Many people associate infrastructure with technology, but that notion is misguided. Technology comes and goes, but infrastructure stays because infrastructure is all about people. Take your favorite moments in life and think about how they were framed by infrastructure. Maybe it is when you got married, and all your loved ones were able to attend the celebration (whether in person or remotely). Or maybe it is when your children were born, and

---

1. Track 9 on the 2013 album *Random Access Memories*.

you needed the heat and light provided by infrastructure more than ever, or later when they played with sprinklers in the summer thanks to water infrastructure. Even if your favorite memory is the time you climbed a mountain or spent a weekend camping, you required transport infrastructure to get there. All our lives are framed by infrastructure.

By now, we are equipped with the engineering principles that govern infrastructure. It is mind-boggling how our most crucial systems rely on relatively few and simple principles that can be explained in one relatively short book. We illustrated how they work in practice by traveling to more than thirty cities in the world and into space over sixteen chapters.

By traveling to Rome, Tel Aviv, Hong Kong, Paris, and Copenhagen, we learned about the world of water and wastewater. We learned that mastering gravity allows us to move and collect water and wastewater.

By traveling to Los Angeles, London, Amsterdam, and Shanghai, we learned about the world of transport. From three relatively simple concepts—flow, speed, and density—we can explain how traffic works and why roads get congested.

By traveling to Cape Town and Chicago, we learned about the world of electricity. We learned that electrical energy is not widely available in raw form and that other forms of energy must be harvested and converted to electricity.

By traveling to Buenos Aires, we learned about the world of natural gas. We learned how natural gas is collected and how processes of adsorption are used to purify the gas and then how that gas is transported through pipelines.

By traveling to Tokyo, we learned about the world of solid waste management and how intricate it is. We learned about the different types of waste and all of the ways to process them.

By traveling to New York City and San Francisco, we learned about the world of telecommunication, from analog telephone systems to wireless and then digital communication.

Finally, by traveling to Singapore, we came full circle and used all of the principles we had learned.

Along the way, we made brief stops in Tokyo (water collection), Bern (water treatment), Hanoi (water distribution), Venice (wastewater treatment), Kuala Lumpur (stormwater management), Boston and Seoul (traffic and roads), Bogotá (public transport), Kyoto (active transport), Lyon (integrated transport), China (Three Gorges Dam; electricity generation), Bohol (electricity distribution), Brussels (natural gas), Cairo (solid waste management), Oaxaca (analog telecommunication), space (digital telecommunication), and Saint Pierre (all infrastructure).

In the end, we covered a lot in this book. My hope is that you will be mesmerized by the infrastructure that surrounds you, just as I am, and that you will keep wanting to learn more.

Next time you go out or travel, look around, try to figure out what that piece of infrastructure is and how it works. Or, even better, visit your local utilities, the sites and facilities that house infrastructure, like your water treatment plant or power plant, and ask questions of the people who work there. I have spoken with many utility providers. Not only are they knowledgeable about the systems they operate, but they are also proud of what they do. I have no doubt they will be happy to show you around and answer your questions.

When you travel around now, can you spot power lines and the three wires? Can you spot wireless antennas? One exercise that I love when I fly is to look out the window, especially during takeoff and landing, and try to identify infrastructure. Often I can spot freight hubs, train lines, and expressways. I can see large transmission power lines. Sometimes I can see power plants, dams, and wind farms. I can also spot which buildings have solar panels installed on their roofs and which have green roofs, as well as those that do not have anything. I also often see large, circle-shaped basins, full of water: these are primary clarifiers used to treat wastewater. Also look at the shape of ponds or small lakes—if the geometry does not look natural, it probably means it is a reservoir. This world of infrastructure that humans have built is amazing and full of surprises.

Knowing what we do now, what about the future of infrastructure? Predicting the future of something as grand and ubiquitous as infrastructure is a thorny exercise. No matter how right we think we are, we will always be wrong, especially if we look in the long term. Some of our predictions may also be wishful thinking or fears, but we can give it a try. First, let's look at major drivers of change that are happening, which should influence how infrastructure will evolve. Second, we can focus on each infrastructure system and list some of the possible ways they may evolve individually.

In terms of major drivers of change, I see at least three: sustainability, resilience, and equity. We have already discussed sustainability a lot. Namely, if we cannot sustain an activity forever, then we should stop it. The most important trend happening in response to the sustainability driver is decarbonization. Decarbonization is primarily about using renewable sources of energy and stopping the excessive emissions of greenhouse gases. Basically, it is primarily about phasing out fossil fuels, especially the ones we are burning. We can consider that some uses of fossil fuels are more noble than others, like making plastic components that are ubiquitous (e.g., in medicine) and can ideally be recycled. In contrast, burning fossil fuels to generate electricity or to move a vehicle is not noble. Once burned, it is gone forever. Decarbonization will require rethinking how infrastructure is planned, designed, and operated, but the goal is clear. At this moment, all infrastructure systems rely on fossil fuels. All will need to change as they decarbonize.

Resilience is another driver of change. Basically, because our lives depend on infrastructure, we need to make sure it works well. But, like everything else, infrastructure is vulnerable. Whether it is due to an extreme event or to normal wear over years, infrastructure breaks and falls apart. In the twentieth century, humans built tons of infrastructure. One article published in 2020 found that the mass of human-made stuff exceeds the mass of all living biomass.[2] In other words, all the stuff the

---

2. E. Elhacham, L. Ben-Uri, J. Grozovski, et al., "Global Human-Made Mass Exceeds All Living Biomass," *Nature* 588 (2020): 442–44.

humans built (primarily infrastructure) weighs more than all the elephants, whales, dogs, trees, plants, ants, fungi, bacteria, and every other living biomass put together. In 1900, the mass of human-made stuff was around 3 percent of all living biomass. All this infrastructure that was built over the past century is bound to deteriorate and fail. Trying to ensure that it functions has become a major driver of change for infrastructure planning and design. This point is further exacerbated by the fact that infrastructure is highly interdependent and interconnected. On purpose, in this book, I looked at infrastructure individually to explain how each system works, but none work in isolation. Water treatment and distribution require electricity and access to roads. Transport infrastructure is impacted by stormwater and relies on electricity and telecommunication. Many electricity-generation methods require a lot of water. And we could go on and on with examples. In 2017, I published a scientific article titled "Urban Infrastructure Is Not a Tree: Integrating and Decentralizing Urban Infrastructure Systems."[3] The premise of the article is that infrastructure systems are not isolated, and yet we tend to plan, design, and build them as if they were. This practice has led us to design infrastructure that is vulnerable since one type of infrastructure breaking down can impact all others. The solution is not to decouple them, since they all benefit from one another, but to integrate them better so that they help one another as opposed to making one another more vulnerable. But building resilient infrastructure is challenging. While the road to decarbonization is clear, the one to resilience is not. Infrastructure systems are complex, and predicting how they respond to shocks and stresses is not obvious.

The third driver of change is equity. Equity is about ensuring that everyone has access to the infrastructure. This point brings us back to the central goal of infrastructure: to provide for people. In the twentieth century, a lot of infrastructure was built for some populations, sometimes

---

3. S. Derrible, "Urban Infrastructure Is Not a Tree: Integrating and Decentralizing Urban Infrastructure Systems," *Environment and Planning B: Urban Analytics and City Science* 44, no. 3 (2017): 553–69.

at the expense of others. Environmental justice has become an important aspect of equity. As we have seen, putting a lot of people together inevitably leads to pollution. Environmental justice is about ensuring that the burdens of pollution are evenly distributed so that we are all responsible for what we generate. Yet marginalized populations often end up bearing an unjustly higher share of pollution, and we need to change that. Hopefully, the quest for sustainability will help, too, as sustainable activities should also result in less pollution.

So far we have assumed that infrastructure is always positive and always desirable, but is it the case? Humans have built an amazing amount of infrastructure. Our society has never been so technologically advanced, and yet are we much happier? In some countries like the United States, life expectancy is getting shorter. Can this development be linked to the absence of infrastructure or maybe to the excess of infrastructure? Often, more infrastructure leads to more pollution and to higher maintenance needs. I recently watched a miniseries on the blue zones—places in the world where people tend to live longer and be healthier. None were incredibly technologically advanced with abundant infrastructure, but all were places where locals had an important social life. Chicago is not a blue zone, but it has a story to tell on this topic. In 1995, Chicago suffered from a massive heat wave that led to at least 739 heat-related deaths. So many people died that bodies had to be kept in refrigerated trucks, like what happened in some cities during the COVID-19 pandemic. After the heat wave, the city took measures to try to prevent a similar tragedy in the future by providing infrastructure solutions. What the American sociologist Eric Klinenberg found in his 2002 book *Heat Wave: A Social Autopsy of Disaster in Chicago* was that infrastructure was largely not to be blamed. Many of the people who died were home alone with no social support. Ultimately, many deaths could have been prevented if someone had checked up on them.

It's not that we do not need more infrastructure; it is that we need better infrastructure, infrastructure that supports and reinforces a stronger social fabric. Loneliness has become a major issue. Equity has the potential to revolutionize the way we plan and design infrastructure, because it will put back at the center the reason why infrastructure is there in the first place: people. I think this is a point many engineers have forgotten. I will repeat what I wrote earlier: infrastructure is not about technology. It is about people. Most times, when I see visions of future cities depicted by companies and organizations, I see flying cars with no traffic congestion and self-regulating infrastructure on a nice sunny day, but I rarely see people. Who are we building this infrastructure for? Rethinking infrastructure to support a strong social fabric is challenging and far from straightforward, but it is critical.

Put in the context of these three drivers of change and in the context of the challenges discussed in each chapter, we can start thinking about how individual infrastructure systems will evolve. I like to say that cities are shaped by the challenges they face. We can therefore recall the challenges discussed in each chapter and think about how addressing them can drive changes in infrastructure.

With water infrastructure, conservation has been a major driver of change in areas that do not have much water to begin with, such as Israel and Singapore. While water conservation efforts will remain important worldwide, I also think that water distribution will need to change in the future. In most cities, water conduits in streets are under high pressure constantly to ensure that possibly contaminated groundwater does not infiltrate the system. This is a large ask that is not resilient, as we are continuously at the mercy of water main breaks. To be able to lower water pressure, we will likely need to start treating water in buildings directly, at the point of use, as is done in Hanoi. In the foreseeable future, water will still need to be treated in water treatment plants, since different sources of raw water require different treatment processes, and it will still need to be distributed, since water must reach buildings, but a final treatment process will likely be applied right before the water is consumed so that we do not rely on perfectly operating water distribution systems. To some extent, the popularity of at-home water filters like those made by Brita® and the significant growth in the water bottle market are symptoms of this need to change how water is distributed.

With wastewater infrastructure, I do not have any predictions about the future of wastewater treatment. To me, the largest change will be in stormwater management. As flooding has become a major issue in virtually every city in the world, green infrastructure and low-impact development can be deployed everywhere. One of the benefits of green infrastructure is that it contributes to making cities more sustainable, resilient, and equitable—sustainable because less energy is required to handle stormwater (compared to deep tunnels, for example), resilient because they make cities less prone to flooding,[4] and equitable because green infrastructure comes with a slew of benefits, from increasing the well-being of people and promoting economic development to lowering ambient air temperature and treating stormwater. To me, green infrastructure is the perfect example of infrastructure that provides for people while supporting the social fabric. My favorite piece of infrastructure in the world is the Cheonggyecheon River in Seoul. If you ever find yourself in Seoul, be sure to visit it, especially during the weekend when many people enjoy it, unless it is raining and is closed to handle stormwater.

With transport, many will immediately think of electric and autonomous vehicles when asked to describe its future, but my mind goes straight to land use. Electric and autonomous vehicles are just technologies. They should not drive the change, but they should support it. Deploying technologies for the sake of doing so makes no sense. We need to deploy technology to improve people's lives. Therefore, we first need to remind ourselves that the sole purpose of transport infrastructure is to enable someone or something to travel from an origin to a destination. That is what I think about. How do we make the destination closer and better? How do we make the journey more comfortable and affordable? And how can this infrastructure support the social fabric of our society? This is where streets become key. As I have argued in the transport chapters, with a car-centric culture, it is like we have been telling people to avoid streets

---

4. Also in a decentralized way, so that we do not rely on one or few systems to manage all the stormwater.

because they are dangerous. It is like we are telling people to shelter themselves or avoid other people at all costs. But what we need as human beings is the exact opposite. We need lively streets with people. That is why my mind goes to land use first. Second, I think of public transport. We have already established why public transport is better than private transport (i.e., physical space). But to me, the current model of public transport is not adapted to current lifestyles. Whether we use current modes (like buses and trains) or new modes (like bike shares), riding transit could improve in many cities. Here we can speculate that autonomous vehicles may allow for smaller and more frequent transit vehicles to work either on fixed routes or on more flexible routes. More generally, we should think integrated transport so that using any mode, or a combination of modes, becomes easier, and we do not have to commit to one mode only. The future of transport is people having access to what they need in a short, affordable, and comfortable way, combined with lively streets that help build the social fabric.

With electricity, the future likely holds more of it. I doubt that our appetite for electricity (and energy in general) will decrease. Plus, often, decarbonization is synonymous with electrification. Fossil fuel–powered vehicles will become electric. Gas furnaces will become electric heat pumps. Gas-powered pumps for water distribution will become electric. The key will be to ensure that all this electricity is generated from sustainable, resilient, and equitable sources, which means more renewable energy and more electricity storage. It also means that the grid will need to be retrofitted to accommodate this new form of electricity exchange (from one-to-many to many-to-many). One of the big uncertainties on the horizon is nuclear fusion. As of this writing, all nuclear power plants use nuclear fission, which works well but generates radiotoxic waste. Nuclear fusion generates vast amounts of clean energy. Having nuclear fusion reactors as part of a diverse energy mix (i.e., generating electricity in several different ways) would greatly contribute in terms of sustainability and resilience. Looking further into the future, wireless electricity transmission may become possible (even maybe from space), but we are not there yet. We should also not be blinded by the possibility of a new technology to the detriment of focusing on people and supporting a strong social fabric, as I have kept emphasizing.

With gas, by contrast, the future holds less of it. Ultimately, the extraction of gas from underground reservoirs will need to stop. We simply cannot rely on a source of energy that is finite, that lowers air quality, and that contributes to climate change. That said, we may still produce some gas from anaerobic digestion (as seen in the solid waste chapter), and I can see small gas distribution networks remaining. Gas can also be easily stored and used for cooking, especially in remote areas. To me, thanks to the knowledge they have acquired, many gas companies could shift their attention toward hydrogen production and transport—a burgeoning energy solution. It is just not common yet, but it shows some promise. Hydrogen reacts with oxygen to create water, and the reaction emits heat. Hydrogen cars exist already. Space rockets are powered by hydrogen. Hydrogen fuel cells can be used to heat buildings. Hydrogen is also useful in agriculture and in various industrial applications. In transport, many see hydrogen as a great energy substitute for long-distance travel, such as for trucks and container ships, in which electricity may be limited. The main problem is in the production of hydrogen. Electrolysis to separate the two hydrogen atoms from the oxygen atom requires energy. The hope is to use clean electricity to drive the process. That brings us back to the future of electricity and the need to generate more of it cleanly.

For the future of solid waste, I see continued efforts on the current trends. We can follow the five Rs. The two strategies of Reduce and Reuse go hand in hand. They involve a shift in the consumer culture, a move away from buying cheap, low-quality, disposable products to buying less affordable, high-quality, long-lasting products. To me, the disposable culture is only a blip in human history. Until the 1950s, most people bought high-quality, long-lasting products, and I think we will return to this tendency. The current trend in electronics of buying more expensive and more reliable products but keeping them longer seems to suggest that. Also, more people now like to spend money on experiences rather than material things. In fact, some research finds that people are happier when they do this. One scientific article mentions that it is in part because "experiential purchases enhance social relations more readily and effectively than material goods."[5] Now we are back to the role of infrastructure supporting a strong social fabric. Still, much solid waste will continue to be generated, and we need to move to Recycling. Quite simply, we should become really good at recycling virtually everything we consume. Think of what I wrote earlier. The mass of all human-made stuff is heavier than all living biomass in the world. We have extracted so much from the Earth already that I would think it will soon become cheaper to recycle what we have extracted than to extract more stuff. Recycling will probably become more predominant once the world population starts to decrease before the end of the twenty-first century. I do not know whether the Recover (i.e., often meaning burning) and Rot (i.e., landfilling) strategies will change much, but their share should decrease in the future.

The last infrastructure we covered was telecommunications. As mentioned, this is the infrastructure I know the least. By speaking with colleagues, I have found that speed is not much of an issue anymore. The future of telecommunication infrastructure is likely the mass deployment of fiber-optic cables everywhere so that everyone and everything has ubiquitous access to the Web—everything seems to be connected now, from watches to refrigerators. I imagine that a lot of the evolution in telecommunication infrastructure will also require electricity (hence the need for clean electricity). I also imagine that many existing protocols will need to be updated. My only concern is how we ensure that this mass deployment contributes to a stronger social fabric. In particular, the verdict on social media that was enabled by telecommunications infrastructure has yet to be handed down. Has social media strengthened or weakened the social fabric? Whatever the future of telecommunications will be, again,

---

5. T. Gilovich, A. Kumar, and L. Jampol, "A Wonderful Life: Experiential Consumption and the Pursuit of Happiness," *Journal of Consumer Psychology* 25, no. 1 (2015): 142–65.

we should not seek to deploy a technology for the sake of it but because it provides for people and helps build a stronger social fabric.

Many of the evolutions that are needed to make infrastructure more sustainable, resilient, and equitable are challenging. When I attend engineering events, I often find the odd engineer who looks defeated, who claims that people do not know how something works and that it—name any infrastructure service—is not easy. I am always puzzled by those people. Engineers solve problems. That is what we are trained to do. That is what we devote our lives to. How can an engineer be so pessimistic with the state of the world or with human ingenuity? No matter how hard, a challenge should always be exciting to solve. An engineering principle should always be exciting to apply or master. Personally, when I get negative, especially when I see corrupted people doing wrong things and promoting wrong ideas, I remind myself that we are all in this together. I recall something that George Costanza says on *Seinfeld*: "You know, we're living in a society!" It always gives me a laugh and puts me in a good mood. Achieving perfection and everlasting happiness is not the goal. Going through it together and aspiring for a better world together, despite our differences, is.

Before we say our goodbyes, I would like to end with two quotes I like a lot. The first is from the Scottish geographer and sociologist, Patrick Geddes. When we think of a city, we usually have an image that pops into our heads, something like a snapshot, often a skyline, that is fixed in time. Patrick Geddes sees it differently: "A city is more than a place in space; it is a drama in time."[6] I like this quote a lot. Like infrastructure, cities are not stationary. They are not in equilibrium. They constantly evolve. And, like cities, infrastructure is a drama in time. Another favorite quote is from Jane Jacobs, whom we met in the active transport chapter in Amsterdam. She wrote, "Cities have the capability of providing something for everybody, only because, and only when, they are created by everybody."[7]

---

6. P. Geddes, "Civics: As Applied Sociology," *Sociological Papers* 1 (1904): 103–38.
7. J. Jacobs, *The Death and Life of Great American Cities* (New York: Random House, 1961), 238.

Become an actor of your city. We have come to depend on infrastructure for virtually everything we do. We have established a symbiotic relationship with infrastructure. Don't only be a consumer of infrastructure; help define it. Infrastructure must be built by people and for people.

Reading this book is only the start of your journey into the realm of infrastructure. What is certain is that infrastructure will evolve, and we should all have a say in how it does. This journey you have decided to embark on is only the beginning, and I promise it will be exciting.

Acknowledgments

First, I want to thank you, the reader, for your curiosity about the built environment surrounding you. Benefiting from infrastructure has become so common and seamless that most people don't even think about it, but you did. You were curious. You wanted to know more and decided that *The Infrastructure Book* would be part of your journey. Thank you.

As with all books, many people have contributed to this book project in one way or another. I thank my literary agent, Joelle Delbourgo, and my editor, Jake Bonar, for their feedback, for seeing how hungry people were to learn about infrastructure, and for believing in this project. I also thank everyone at Prometheus Books for making the publication process easy and exciting.

This book contains quite a bit of technical content. While I have gained expertise in the seven infrastructure systems featured in these pages, I must thank many people for providing helpful information or for checking some of the technical content. Some are colleagues or people I talked to at some point and asked questions. Others are friends or even friends of friends with whom I have barely interacted but who volunteered to read part of the book or talk to me. Still others are utility managers and professionals who are passionate and dedicated to making the infrastructure they oversee function and who had the patience to answer my unrelenting questions.

In no particular order, I thank Christopher Kennedy, Eugene Mohareb, Mikhail Chester, Matthew Eckelman, Aslihan Karatas, Joseph Schulenberg, Thomas Theis, Joel Rogers, Nicolas Picard, Bat Hen Nahmias Biran, Nadav Biran, Amid Khodadoust, Pierre Mayol, Shauhrat

Chopra, David von Eiff, Andrew Billing, Caitlin Cottrill, Francisco Camara Pereira, Jane Lin, Abolfazl (Kouros) Mohammadian, Paul Hoekman, Mariella Uchida, Florencia Gonzalez Otharan, William (Bill) Ryan, Edward Oughton, Erik Boch, Julius Kusuma, Lynette Cheah, Lih Wei Yeow, Mohit Arora, Yu Dong, Nikita Patel, Martin Melosi, and Juan Acosta Sequeda. I am sure I have missed some people.

I must also thank all the students who took the general education class "CME112: Evolution of Infrastructure and Society" at the University of Illinois Chicago, which adopted drafts of this book before it was published. Thank you for your positive feedback. More generally, thank you to everyone who has commented on this book, answered my never-ending questions, or simply encouraged me along the way.

I started *The Infrastructure Book* in spring 2020 during a sabbatical in Paris. I had published my textbook *Urban Engineering for Sustainability* (2019) and yearned to turn it into a popular science book to explain to a larger audience how infrastructure works. Little did I know that a global pandemic would force me to remain indoors, giving me a higher appreciation for all the infrastructure I had access to from the comfort of the apartment where I stayed. Thank you to Thomas and Charline Petit for offering their apartment in Paris to us while they were away. It provided the perfect setting to start this book.

Once back in Chicago, I continued to work on the book whenever time allowed, between research projects, scientific articles, and classes. At a time when traveling physically was impossible, I traveled virtually to all the cities featured in the book. Making time to work on a book can be challenging, but writing *The Infrastructure Book* became my escape. To unwind, some people go to the gym or watch TV; I write. For someone who despised writing as a teenager, I have come around big time.

An enormous thank you goes to my family for their unyielding support for whatever crazy task I set for myself, especially to my father, who has always been a role model and a constant and continuous source of support no matter what happens in our lives.

The biggest thank you of all goes to my wife, Marie-Agathe. This book is dedicated to her. Marie-Agathe is my partner in life and travel. We have crossed the world together and lived in five countries over three continents. Since I have moved so much, no physical place feels like home. When people ask me where I feel most at home, I reply, "Wherever Marie is." She's my everything. I am the luckiest man in the world. Marie: Je t'aime plus que tout. Merci.